Lessons Learned Troubleshooting Hydronic Heat

Ray Wohlfarth

This book does not take the place of the boiler manufacturers written instructions, engineering, or code issues that may be in force in your locale. Please follow the boiler manufacturer's instructions included in the boiler installation manual. This does not take the place of a properly designed system from an experienced designer. Thank you for choosing to purchase and read this book.

Copyright © 2024 Ray Wohlfarth

All rights reserved.

DEDICATION

This is dedicated to my amazing family: Sheila, Jon, Conor & Lyndsay, Abby & Mike & their children, Delaney & Owen, Samantha & Ryan & their children, Jake & Annie. I also must acknowledge my friend who started my on this path, Dan Holohan. Thanks for being in my corner.

Table of Contents

- THE BOILER IS NOT FIRING 6
- PILOT LIGHTS BUT MAIN FLAME DOESN'T 7
- INTERMITTENT FLAME FAILURES 7
- BURNER SHORT CYCLES 8
- THE BOILER IS OFF ON LOW WATER 8
- BURNER BLOWER STARTS BUT FLAME DOESN'T 8
- WATER UNDER BOILER 9
- BOILER PRESSURE KEEPS RISING 9
- LOW WATER TREATMENT CHEMICAL LEVEL 9
- DISCOLORATION OF BOILER JACKET 9
- HIGH FUEL COSTS 10
- FLAME ROLLS FROM UNDER BOILER BASE 10
- THE BOILER IS SOOTED 10
- THE TEMPERATURE RISE THROUGH THE BOILER IS TOO LOW .. 11
- THE FLUE GAS TEMPERATURE IS HIGHER 11
- MANUAL RESET LIMIT CONTROL IS TRIPPING 11
- THE TEMPERATURE RISE THROUGH THE BOILER IS TOO HIGH .. 11
- ENTIRE BUILDING IS COOL 13
- SOME PARTS OF THE BUILDING ARE COLD 13
- ONE AREA OVERHEATING 13
- TOP FLOOR COOL WHILE BOTTOM FLOORS ARE WARM ... 14
- RADIATOR NOT HEATING 14
- BEARING ASSEMBLY IS LEAKING 14
- UNBALANCED HEAT IN BUILDING 14
- NO HEAT TO AHU COIL 14
- EXCESSIVE AIR IN SYSTEM 14
- SYSTEM IS LOSING WATER 15
- CORROSION IN SYSTEM 15
- RADIATOR IS DRIPPING 15
- NO HEAT FROM BASEBOARD RADIATION 15
- THE EXPANSION TANK KEEPS FLOODING 16
- HOW TO TELL IF EXPANSION TANK IS FLOODED 17
- RELIEF VALVE IS OPENING 17
- BOILER NOISES 22
- SQUEALING NOISE WHEN BOILER RUNS 22
- BOILER MAKES A MOANING SOUND WHEN FIRING 22
- THE BOILER MAKES A TICKING / CRACKING NOISE WHEN FIRING 22
- THE BOILER MAKES A HUMMING OR BUZZING SOUND ... 22
- THE BOILER MAKES A HISSING SOUND WHEN FIRING 22
- THE PUMP IS SQUEALING 22
- CHATTERING IN PIPES 22
- PIPING MAKES A GURGLING NOISE 22
- EXCESSIVE COMBUSTION NOISE 23
- MAIN AIR VENT IS SQUEALING / NOISY 23
- PIPES ARE BANGING 23
- PUMP IS GRINDING / BANGING 23
- EXCESSIVE VIBRATION IN SYSTEM 24
- BASEBOARD RADIATION IS TICKING 24
- COMMON HYDRONIC PIPING 24
- TROUBLESHOOTING HYDRONIC PIPING 29
- TROUBLE-SHOOTING HYDRONIC SYSTEMS 31
- ZONING WITH PUMPS 34
- ZONING WITH VALVES 35
- AIR VENTING TIPS I LEARNED 37
- UNDERSTANDING THE BOILER SEQUENCE OF OPERATION 39
- DEGREE DAYS AND DESIGN TEMPERATURES 42
- OUTDOOR DESIGN TEMPERATURES CANADA 45
- HEATING FORMULAS 45
- EXPANSION COMPRESSION TANK SIZING 47
- PUMPS 49
- ESTIMATE HYDRONIC SYSTEM VOLUME 52
- GLYCOL 53
- VELOCITY CALCULATION 56
- CARBON MONOXIDE DANGERS 59
- TYPICAL COMBUSTION READINGS FOR CATEGORY I BOILERS .. 59
- CLOCKING A GAS METER 60
- WATER TREATMENT 63
- COMBUSTION AIR 66
- LEAK TESTING GAS VALVES 71
- SIZING GAS TRAIN MANIFOLD VENT 71

Boiler Room Safety

A boiler room or basement can be a dangerous place. Be sure you are working in a safe environment. The following are suggestions I have for a safe service call.

Plan your escape – If the room fills with steam, it will have no visibility and steam displaces oxygen. Look around  for any trip hazards if you must leave in a hurry. I know what you're thinking, I'm not working on a steam boiler. Both a hydronic boiler or a water heater could generate steam if the temperature is high enough and the relief valve opens.

Shut off boilers until you are sure the room is safe.

Is there carbon monoxide in the room? I suggest you use a personal CO detector when entering.

Is there combustion air? Is it large enough? Is it blocked?

Do you smell natural gas or sour flue gases. This should be corrected right away.

Is the flue for the water heater and boiler intact, properly pitched, and safe?

Is the boiler or water heater jacket discolored or has black streaks? This could indicate carbon monoxide forming.

If your boiler room was flooded and the boiler or burner was under water, do NOT operate the boiler until it is checked by an expert. Consult the boiler manufacturer as they most likely will tell you to replace the boiler.

Is there rust under the water heater or boiler draft diverter? This could indicate back drafting.

Is there anything dangerous stored in the boiler room? Gasoline, lawn mower, or snow blower. The following a list compiled by Weil McLain of things to avoid storing in a boiler room.

What is stored in boiler room? Some items do not react well with flames and could create a hazardous situation. The following is a list of items Weil McLain says should be avoided in a boiler room:

Products to avoid in a boiler room according to Weil McLain
Spray cans containing chloro/fluorocarbons
Permanent wave solutions
Chlorinated waxes/cleaners
Chloring based swimming pool chemicals
Calcium chloride used for thawing
Sodium chloride used for water softening
Refrigerant leaks
Paint or varnish removers
Hydrochloric /Muriatic acids
Cements and Glues
Antistatic fabric softeners used in clothes dryers
Chloring type bleaches, detergents, and cleaning solvents found in household laundry rooms
Adhesives used to fasten building products and other similar products

Ever since December 1899, hydronic systems were designed using $180^0 F$ as the design supply water temperature.

How radiator covers affect heat output

The heat output from a cast iron radiator is 60% convection & 40% radiation. When you cover a radiator, it prevents convection & the heat output is reduced.

5% loss of capacity
(2")

20% less capacity
A* — 3"
* If A is 50% of radiator width, it loses 10% capacity.
* If A is 150% of radiator width, it loses 35% capacity.
2"

30% loss of capacity
A
2"   2"

Types of Heat

Sensible Heat	Sensible heat is any heat transfer that causes a change in temperature without causing a change of state. Sensible heat can be measured with a dry bulb thermometer.
Latent Heat	Latent heat is the amount of heat required to cause a change of state. In a boiler system, this would be the amount of heat added to water to change it from water to steam. It requires 970 Btus to raise 1 pound of water at $212°F$ to steam.
Total Heat	Total heat is the sum of the sensible and latent heat in an exchange process. It is sometimes called enthalpy.

Heat output from a radiator

60% of the heat from a radiator is by convection and 40% is through radiation

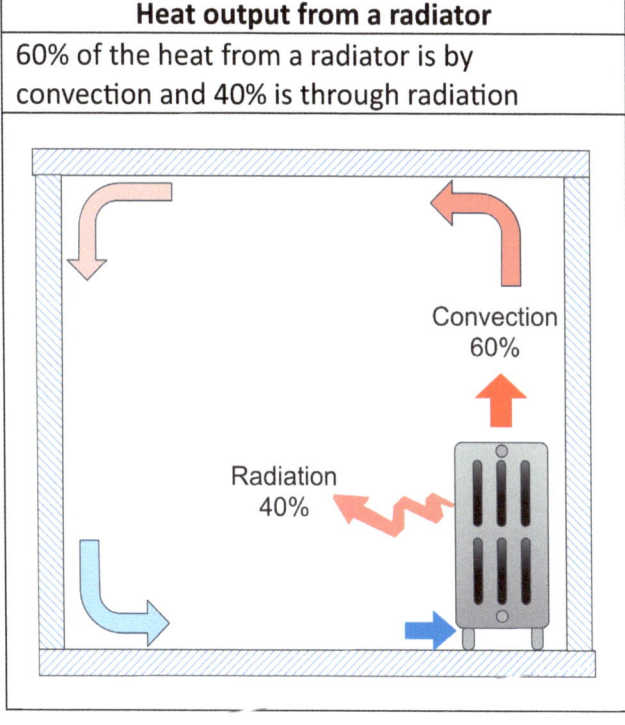

No heat

The boiler is not firing.

Power	If there is no power, check the breaker, switch, manual reset limit control, or the emergency door switch.
Call for heat	Is there a call for heat? Verify building control system is calling for heat.
Water temperature	Check the PTA gauge to see if the boiler is up to temperature.
Fuel	Verify there is gas pressure at the inlet to gas train. If not, look for closed gas valves.
Flame failure	If the boiler is off due on flame failure, press the reset button once. If it trips again, have the burner checked to determine the cause. Do not press reset more than once without checking the burner!
Low water cutoff	Verify water is present in the boiler. Press the manual reset on the low water cutoff.
Gas pressure switch	Gas pressure switches on the gas train assure the gas pressure to the boiler is correct. These controls have a manual reset button.

Gas pressure switch

Limit control	Verify boiler temperature is lower than the limit control setpoint. The boiler limit control is a manual reset control.
Pilot	Verify pilot solenoid valve opens and pilot gas pressure regulator

	has proper gas pressure. Most flame failures occur during the pilot sequence.
Emergency door switch	The door switch located just in/outside the boiler room could be pushed or switched off.
Adjustable bleed valve	The adjustable bleed valve, used to slow the opening of the gas valve, could be closed. This is located on the gas valve. Open valve a half turn and try again.

Adjustable bleed valve

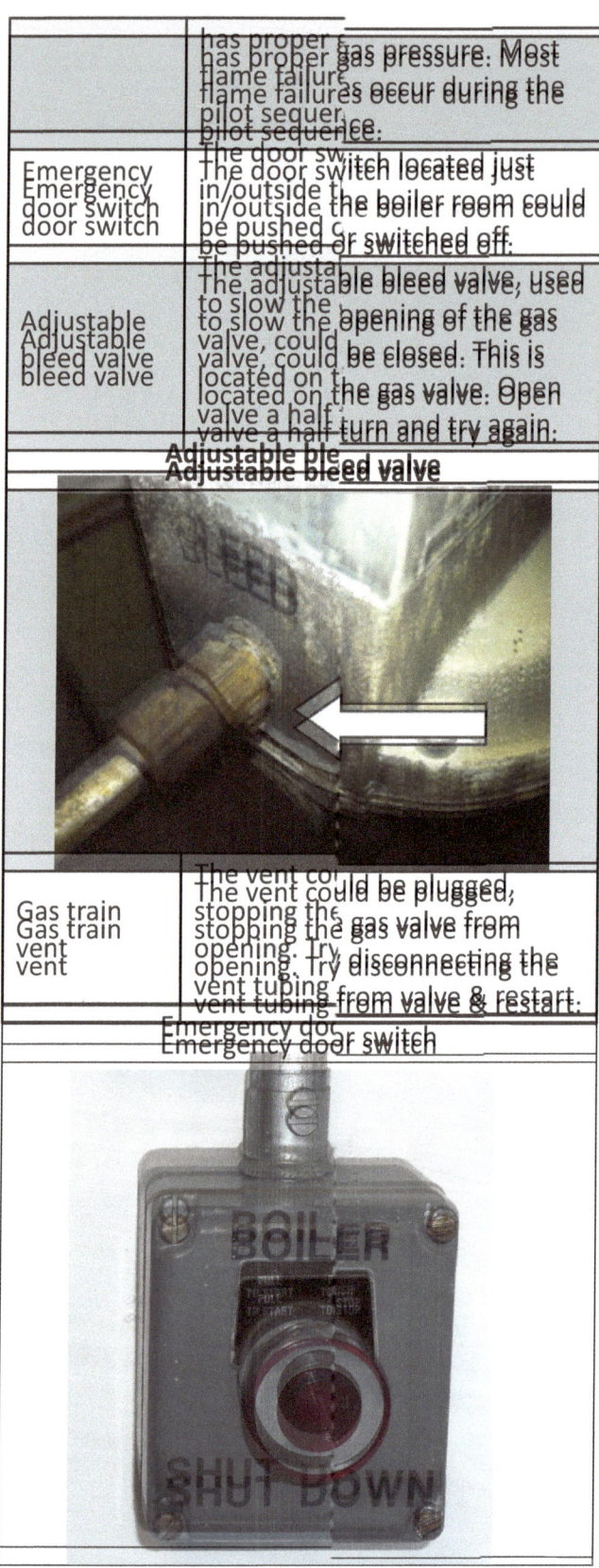

Gas train vent	The vent could be plugged, stopping the gas valve from opening. Try disconnecting the vent tubing from valve & restart.

Emergency door switch

A 1" pipe can carry as many Btus as a 20" round duct.

Pilot lights but main flame doesn't

Gas train vent	The vent from the gas pressure regulator or gas valve could be plugged, stopping the gas valve from opening. Try disconnecting the vent tubing from valve & restart.
Pilot flame not sensed	Verify there is a flame signal during the pilot sequence.
Defective gas valve	Check gas valve operation.
Closed gas valve	Verify the manual gas valve downstream of pilot tapping is open.

Intermittent Flame Failures

Loose wire connections	Verify each wire terminal screw is snug.
Electrical voltage	Verify electrical voltage is within the allowable range.
Pilot solenoid	Verify pilot solenoid valve opens.
Pilot gas pressure regulator	Verify pilot pressure doesn't drop during pilot sequence.
Enable relay is a Triac type relay	Triac relays allow voltage leaks through before relay contacts fully make. Replace with definite purpose relay.
Variable frequency drive or VFD Drive is too close	Install a power line "noise" filter at the supply side of the boiler's power-on switch.
Draft proving switch	Verify draft proving switch is working properly. Check for blockage in draft sensing tube.
Gas pressure	Verify gas pressure does not drop below minimum setting. If there is more than one appliance in the room, check gas pressure when all are operating.

Flame safeguard	A failing flame safeguard control could cause the flame failures.
Pilot flame	A pilot flame can be pulled away from the flame sensor.
Dirty flame sensor	Try cleaning the flame sensor.
Defective flame detector	The ultraviolet or infrared flame detector could be weak and failing.
Flame signal drops	Watch flame signal during pilot and main flame to verify it doesn't drop.
Excess draft	Verify the draft is not too high and pulling the flame away from the sensing rod.
Gas valve actuator leaking	If you see hydraulic fluid in the window, the valve actuator should be replaced. It could cause flame failures and eventually fail.

Leaking gas valve actuator

Look here for hydraulic fluid leak

A blower requires 10 times more energy than a pump.

Burner short cycles	
Flame rod is dirty or misaligned	Check flame signal & clean or replace the flame rod.
Flame detector is defective	Check flame detector for proper flame signal.
Burner air flow switch	Verify the airflow sensing tube is not restricted. Verify the control is not misadjusted or defective.
Gas train vent	Verify the gas train vent is not plugged or restricted.
Chimney draft too high	The draft may be pulling the flame away from the flame sensor.
Temperature control	If the control differential is set too close, the boiler could short cycle.
Insufficient boiler flow	If the water flow through the boiler is too low, it will short cycle. Verify proper water flow rate.
Boiler losing call for heat	Verify the boiler has a call for heat. It could be a thermostat losing the call or another safety control.
Reset control temperature sensor is in wrong location	If the loop temperature sensor is too close to the boiler outlet, it could short cycle.
Boiler piped wrong	If the boiler is piped backward, the supply water may get pulled back into the boiler return.
Low heat load in building	Consider installing a buffer tank

The boiler is off on low water	
No water in system	This indicates a very serious problem, and you should investigate where the water went. Check the tridicator or PTA gauge to verify there is pressure in the system.
Defective low water cutoff control	Verify the low water is working properly.

Burner blower starts but flame doesn't	
Air flow switch	Verify sensing tubes are not kinked or plugged. Check air flow switch.
Igniter	Verify there is a spark. Check electrodes, electrode wiring, ignition transformer.
Adjustable gas valve orifice	Verify the gas valve vent orifice is open.
Plugged gas train vent	Verify vent tubing is not plugged or restricted
Pilot solenoid valve	Verify the pilot solenoid valve operates
Excess combustion air	Too much air could blow the flame away from the sensing rod
Gas pressure	Verify gas pressure does not drop below minimum setting. If there is more than one appliance in the room, check gas pressure when all are operating.
Flame safeguard	A failing flame safeguard control could cause the flame failures.
Pilot flame	A pilot flame can be pulled away from the flame sensor.
Dirty flame sensor	Try cleaning the flame sensor.
Defective flame detector	The ultraviolet or infrared flame detector could be weak and failing.
Defective gas valve actuator	Look for leaking hydraulic fluid in window of valve actuator. If it is, replace the actuator.
Excess draft	Verify the draft is not too high and pulling the flame away from the sensing rod.

Boiler Issues

Water under boiler	
Check boiler for leaks	Shut off the boiler and look at the fireside to see if you see any leaks.
Check piping for leaks	Look around the boiler room for pipe leaks. Repair the leaks ASAP.
Cold water in boiler	When the boiler is starting with room temperature water, water could drip. If the leak continues after the boiler starts to heat, look for a boiler leak.

Boiler draft is wrong	
Boiler draft is low	**Boiler draft is high**
Excessive heat in boiler	Elevated flue temperatures
Flue gas spillage into room	Lowers boiler efficiency
Could allow Carbon Monoxide to form	Could allow Carbon Monoxide to form
Affects boiler efficiency	Flame impingement on metal surfaces
Possible damage to burner	
Flame impingement on burner head	

Boiler pressure keeps rising	
Water feeder	Verify water feeder shuts off at proper pressure. If not replace it.
Compression tank	Verify compression tank is not flooded or piping feeding the tank is not plugged.

Low water treatment chemical level	
System leak	Look for water leaks and repair.
Leaking or defective backflow preventer	The boiler water could be back flowing into the potable water and being dissolved. This is very dangerous as the chemicals may be harmful if ingested.

Discoloration of boiler jacket	
Flue gas leak	If the hot flue gases are not going up the flue, they could cause discoloration of the boiler jacket.
Negative condition in boiler room	An exhaust fan in the boiler room could pull the flue gases from the boiler and into the room.
Insufficient combustion air	If the combustion air opening is blocked or restricted, it could cause the flue gases to roll out of the boiler and into the boiler room.
Plugged flue passages	If the flue passages are plugged, the flue gases could spill from the boiler into the boiler room.
Refractory	Refractory could be missing or damaged
Combustion air	The combustion air may have changed. Look for closed or blocked combustion air openings.

Discoloration of jacket could signify dangerous back drafting of flue gases and boiler should not operate until checked.

Discolored boiler jacket

High fuel costs

Undersized or Underfired boiler	Undersized boiler will run excessively
Burner short cycling	If the burner starts and stops, it uses more energy.
Air vent not working	Air is trapped inside piping and water cannot get in.
Incorrect near boiler piping	Carryover could form due to incorrect near boiler piping.

Rust under draft diverter

Negative conditions	If you see rust under the boiler or the draft diverter, it could be due to a negative condition in the boiler room
Underfired burner	If the burner is not firing to capacity, the flue gases could be condensing.
Blocked flue	Check for flue blockage
Blocked fireside of boiler	Verify the fireside is not plugged with soot or debris.

Flame rolls from under boiler base
(Atmospheric type boiler)

Negative condition	Exhaust fans in the building could pull the boiler room into a negative. -0.03" WC could pull flue gases from boiler or water heater.
Insufficient combustion air	Verify sufficient combustion air

Blocked flue	Check for flue blockage
Blocked fireside of boiler	Verify the fireside is not plugged with soot or debris. Look for boiler leak.
Gas pressure regulator downstream of the electric valve?	Verify gas pressure regulator is upstream of electric gas valves.
Valve opening too fast	Adjust bleed orifice in gas valve

Flame rollout

The boiler is sooted

Negative condition in boiler room	Boiler room could be in negative condition due to exhaust fans in building.
Insufficient combustion air	Verify the room has enough combustion air.
Blocked flue	Check for flue blockage
Water temperature is too low	Low water temperature could cause flue gas condensation & sooting. Non-condensing boilers should operate above 140°F.
Blocked fireside of boiler	Verify the fireside is not plugged with soot or debris.
Incorrect air to fuel adjustment	Check the efficiency of the boiler with an analyzer. Adjust as needed. Verify gas pressure is not too high.
Boiler is leaking	Water leaking from the boiler could cause sooting.

Soot is very dangerous and extreme care required when cleaning it.

The temperature rise through the boiler is too low

Scale	If the water side is scaled, it will impede heat transfer
Fireside dirty	If the fire side is dirty or sooted, it will impede heat transfer
Building doesn't need heat	If the building is warm and doesn't need heat, the delta t is lower.

The flue gas temperature is higher

Scale	Scale could be forming on the water side of the boiler. The water treatment specialist should be notified. Scale impedes the heat transfer in the boiler.
Dirty fireside	If the fireside of the boiler is dirty, this could result in higher flue temperatures and fuel consumption.
Overfired	The burner may be overfiring in the boiler, causing more heat than the boiler can handle.
Excess draft	The draft or velocity of the flue gases traveling through the boiler is too high. Results in insufficient heat transferred to the boiler.

Excess water flow	If the water is traveling too quickly, it won't have time to absorb the heat.

Manual reset limit control is tripping

Temperature setting between the operating and limit control is too close	Manual reset control should be 20-30°F higher than the operating temperature control setpoint.
Defective operating control	If the burner fires above the operating temperature setpoint, it could be defective.
Defective limit control	The limit control may shut off the boiler lower than the setpoint.
Boiler isolation valve	The hot surfaces inside the boiler continue generating heat after the burner shuts off. If an isolation valve closes immediately after the burner shuts off, the temperature inside the boiler keeps rising.

Condensing boiler not condensing

Water temperature is too high	If the water temperature is over 140°F, a condensing boiler will not condense.

The temperature rise through the boiler is too high

Low flow	Insufficient flow could cause a higher delta T
Underfired boiler	If the burner firing rate is low, the delta T is high.

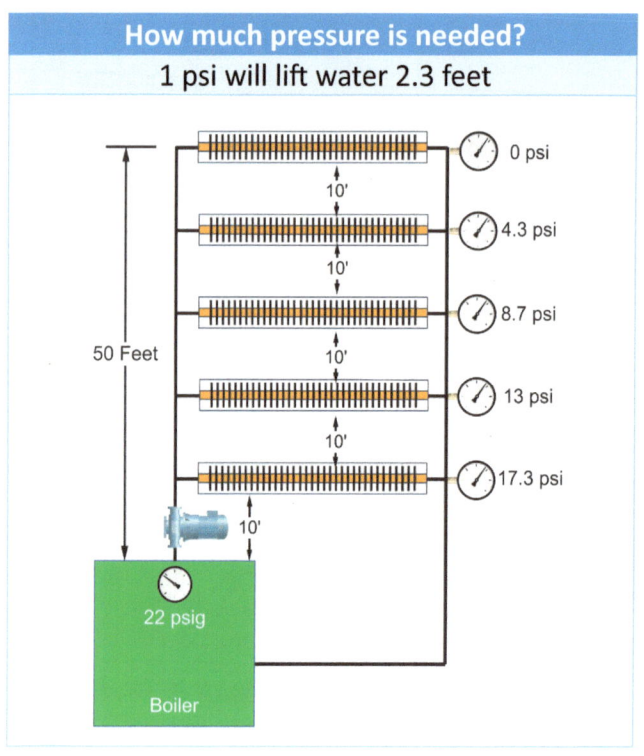

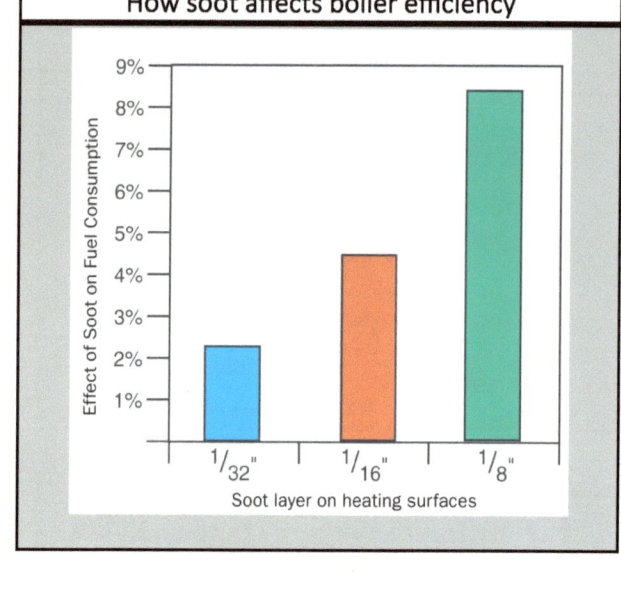

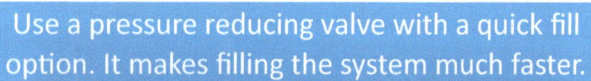

This five-year boiler was destroyed when chilled water was allowed to flow through it.

Typical air velocity over a hydronic coil is 400-500 feet per minute.

Underheating /Overheating

Entire building is cool	
Reset control adjustment	Verify the reset control is adjusted properly.
Water temperature too low	Verify the boiler water temperature is warm enough.
Outdoor air sensor is in sunlight	Verify sun isn't shining on the outdoor temperature sensor, it way be a false reading.
Undersized boiler	If the boiler is too small, it will not heat the building
Thermostat in wrong place	If the thermostat has sun shining on it, it may be reading a warmer temperature.
Pipe check valve	Verify check valve works and is not backward.
No call for heat	Is the call for heat being lost? Check thermostat or thermostatic wiring.
Pump not working	Verify pump is working & in proper direction
Pipes are too small	Verify pipes are large enough for the load.
Water temperature sensor in wrong location	If the temperature is on the supply pipe, it could be short cycling the boilers. Try switching it to the return side.

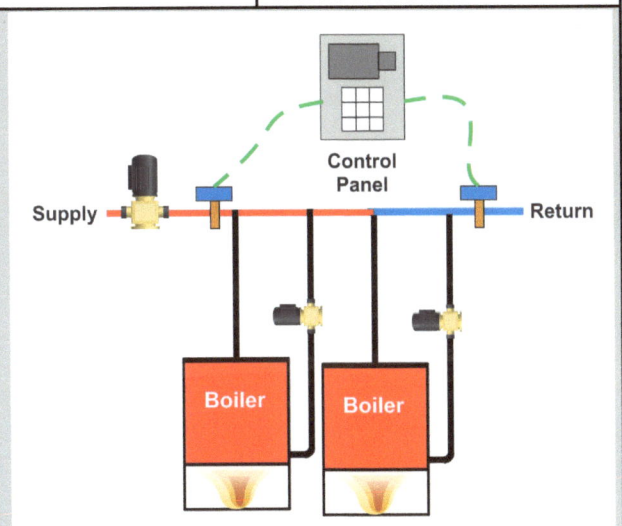

Some parts of the building are cold

Radiators or piping is air bound	Vent radiators or piping to remove air
System pressure too low	1 psi can lift water 2.3 feet. Measure the highest radiator above the boiler, divide by 2 and that should be the pressure on the PTA gauge.
Zone valve / circulator not working.	Verify zone pumps operate and or zone valves open & close.
Missing pipe insulation	Missing pipe insulation causes pipes to cool before reaching the radiator.
Is radiator covered?	Covering the radiator reduces the heat output. Change or remove cover.
Strainer plugged	If the strainer is plugged, water may not get through.
Thermostat not working	Verify thermostat works
Water leak	Check for water leaking from zone pipe.
Pump relay is defective	Verify pump relay works
Different heat emitters	Copper heats & cools quickly while cast radiators heat & cool slowly. Its best to have them on separate zones.

One area overheating

Ghost flow	Verify flow is not going backward through the idle zone.
Leaking valve	Verify zone valves close properly.
Defective thermostat	Verify thermostat works properly.

If you double the size of a pipe, you increase the capacity by 4 times.

Top floor cool while bottom floors are warm

System pressure too low	Verify there is enough pressure to reach the top floor.
Direct return piping system	Flow may be going through the closest radiators. Try closing the radiator valves slightly to get more heat to other floors.
Air in system	Vent air from highest radiators.
System was a gravity distribution	Check for orifices the top radiator valves

Radiator not heating

Radiator is air bound	Bleed air from radiator
No hot water to radiator	Is the supply pipe warm? If not, look for a closed valve in piping
Heat up to radiator, none in radiator	Check radiator valve. The gate may have fallen off the valve.
Return pipe blockage	If the cool water can't get out, hot water can't enter. Look for closed valve or restriction.
Radiator is enclosed	If the radiator is enclosed, the heat may not be able to escape.
System used to be a gravity system	If the system was converted from a gravity system, the water may shoot across the bottom of the radiation. Try closing the valve all the way and opening it a full turn. This may slow the velocity, so the hot water rises.

Bearing assembly is leaking

Defective bearing assembly	Replace bearing assembly

Unbalanced heat in building

System balance	On a direct return system, the rooms closest to the boiler may overheat. Try throttling the flow to the closest rooms to force more into furthest rooms
Zone valve not open	Verify zone valve opens and closes on a call from the thermostat.
Zone circulators not working	Verify the zone circulators operate on a call from the thermostat.
Valve closed	Verify valves are open to the system

No heat to AHU coil

Air in coil	Bleed air from coil
Defective valve	Verify valve operates
Low system pressure	Verify static pressure is high enough to reach the coil.
Check delta T	Verify proper ΔT across the coil on the air & water side.
Low flow	A high delta T could mean low flow to coil.

Excessive air in system

Not pumping away	If the pump is directed to the expansion tank, air bubbles could form.
Air vent not working	Verify air vent works
System leaks	Check for & repair water leaks
System pressure too low	Check static pressure and increase if needed.

Typical duct systems lose 25-40% of their air through duct leaks -US DOE

System is losing water

Water leaking	Check & repair any water leaks. If you can't find the leak, check the air handling unit at the coils. The leak may be there and going down the AHU drain.
Boiler leaking	An internal boiler leak could cause the system to lose water.
Missing or damaged backflow preventer	If the system pressure is higher than the building water pressure, it could back feed into the building's potable water.
System water pressure too high	If the system water pressure is too high, water could leak from relief valve or out the valve packing.

Corrosion in system

System leaking	Check & repair system water leaks
System pH	Check system pH
Improper electrical ground	Verify electrical system is properly grounded
Dissimilar metals	Very dissimilar metals are not touching each other. Use dielectric fittings between them.

Radiator is dripping

Valve packing is loose	Try tightening valve packing nut ¼ turn.
Radiator is leaking	Look for crack or hole in radiator.

No heat from baseboard radiation

Air in piping	Bleed air from piping
Defective valve	Verify valve operates
Low system pressure	Verify static pressure is high enough to reach the baseboard.
Vent louver closed	Verify the louver on top of the baseboard is open.

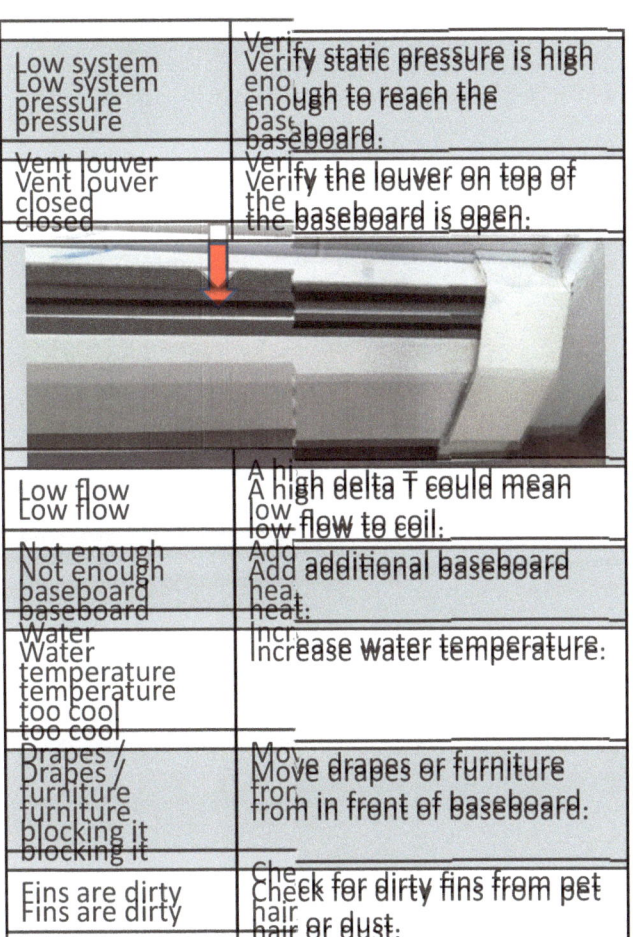

Low flow	A high delta T could mean low flow to coil.
Not enough baseboard	Add additional baseboard heat.
Water temperature too cool	Increase water temperature.
Drapes / furniture blocking it	Move drapes or furniture from in front of baseboard.
Fins are dirty	Check for dirty fins from pet hair or dust.

What is a triple duty valve?

A triple duty valve is a shut off, flow control, and check valve.

Expansion/ Compression Tank

The expansion tank keeps flooding	
Gauge glass is leaking	Check the gauge glass that shows the water level in the tank. The top nut could be leaking. Try tightening the nut a ¼ turn. You may need to replace washer and ring.
There's a hole in the tank	Look on the tank for a hole which may be leaking air from inside.
Pressure reducing valve is defective or misadjusted.	Try closing manual valve in makeup water pipe. If the tank water level continues to rise, it's not the PRV and look at the tank.
Undersized expansion tank	If the tank is undersized, it could flood.

Testing the Air Pressure for the Diaphragm Compression Tank

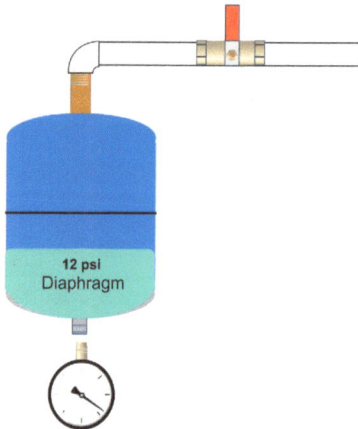

The diaphragm is pre-charged at the factory with 12 psig air pressure. The bladder air pressure should match the hydronic system static pressure. Divide the height of the highest radiator by 2 and that should give you the system pressure. Disconnect the tank from the system water. Use a tire pressure gauge. If the static pressure is higher than the air pressure inside the diaphragm, the air pressure in the tank should be increased. Once the correct pressure is verified on the diaphragm, the tank can be connected to the hydronic system. The bladder will typically lose about one psig per year and should be checked yearly.

Draining a horizontal steel expansion tank
Open brass fitting at bottom of Airtrol fitting and drain until water stops. That should be about 1/3 air cushion. The Airtrol fitting has a tube extended into the tank.

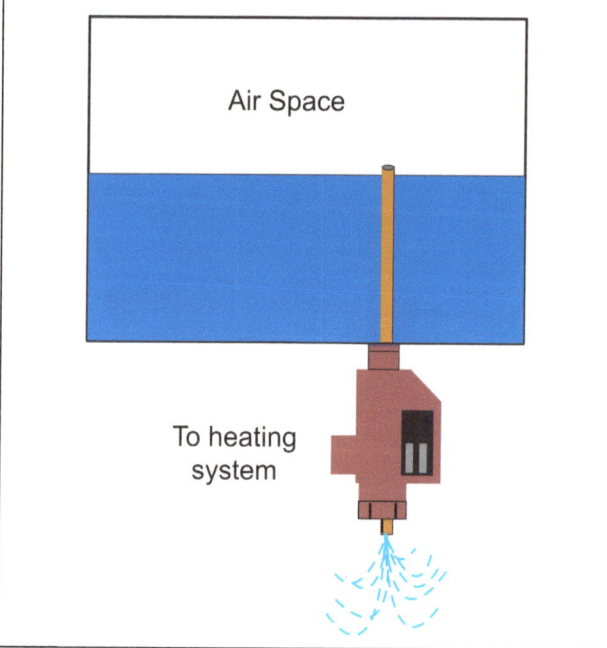

Most commercial buildings need an ASME rated expansion tank. ASME is the American Society of Mechanical Engineers.

How to tell if expansion tank is flooded
Watch pressure dial on tridicator dial. If the pressure rises when the burner fires, this could indicate the expansion tank is flooded.

Expansion tank flooded
Try tightening the top gauge glass nut and the valve packing nut ¼ turn.

Testing for leaking pressure reducing valve
Close valves to backflow preventer and the pressure reducing valve and watch the pressure. It may take a day or so. If the pressure keeps rising after the valves are closed, it may be a blockage on the piping to the expansion tank or the tank is leaking.

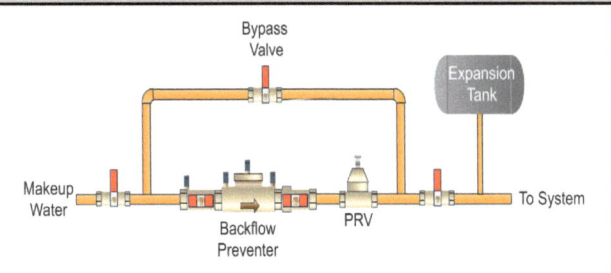

Relief Valve

Relief valve is opening	
System pressure is too high	Check if water feeder shuts off at proper pressure.
Compression tank flooded or too small	Verify tank is not flooded and check sizing.
Defective relief valve	Check condition of relief valve

Relief valve discharge piping

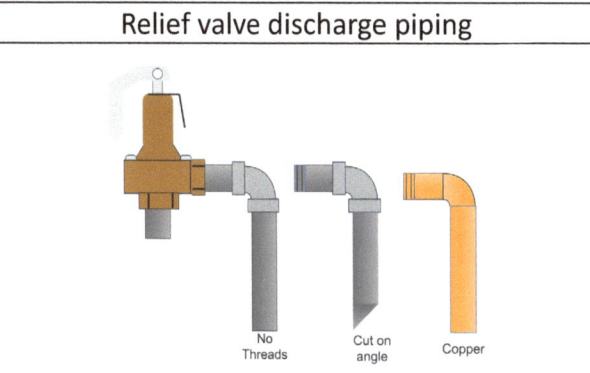

Incorrect way to pipe discharge pipe. Pipe needs room for expansion.

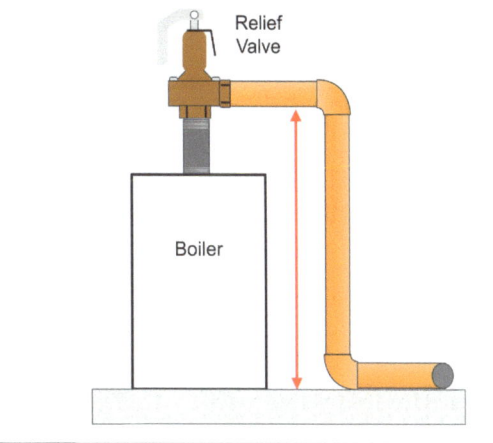

This leaves room for expansion

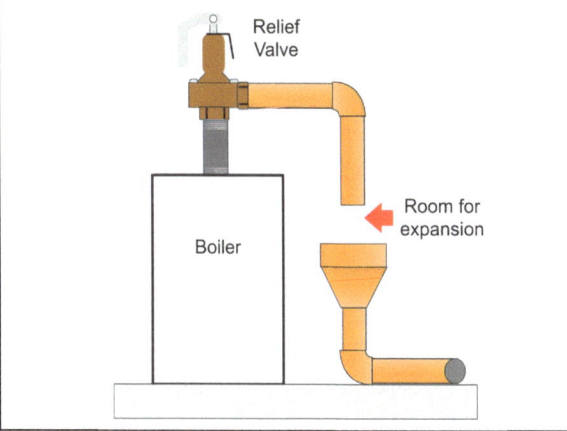

Boiler Temperature Controls

Commercial boilers require a minimum of two temperature controls but may have more. One is an automatic reset which is often called the Operating Control. It is set for the desired water temperature. The other is called a Limit Control, and it's a manual reset control. This is installed in case the operating temperature control fails. If the burner is not an on off burner, it may have a firing rate control. This control varies the burner input according to the water temperature.

Residential boilers typically have one temperature control which is an automatic reset. It is set for the desired water temperature, typically 180°F.

Hydronic temperature controls
Operating control (Left side) is an automatic reset. Usually set for 180°F.
Limit control (Right side) is a manual reset and typically set for 20-30°F higher than the operating control.

> A space heater consumes as much energy as twenty 2-bulb fluorescent fixtures.

Dual Aquastats

Honeywell L8148 series aquastat

This is an automatic reset temperature control. The Red Arrow (HI) is typically set for 180°F. The Blue Arrow (Diff) is the differential and set for 10-15°F. Most of these controls do not have this knob and use a fixed differential. The Green Arrow (LO) may have several different options. Sometimes the low setting is used for the firing rate or the minimum boiler water temperature.

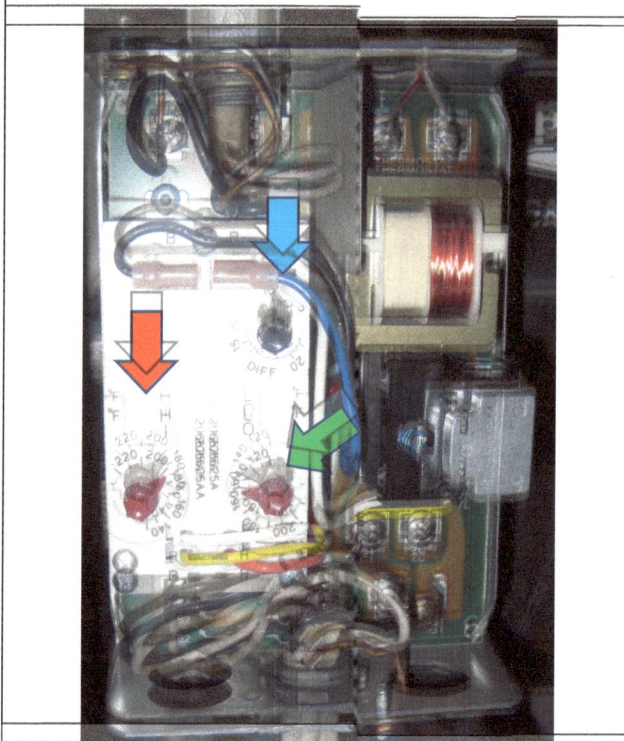

Honeywell L8148 series aquastat

This is an automatic reset control. Usually set for 180°F. See arrow below. In most applications, it controls the burner and circulator.

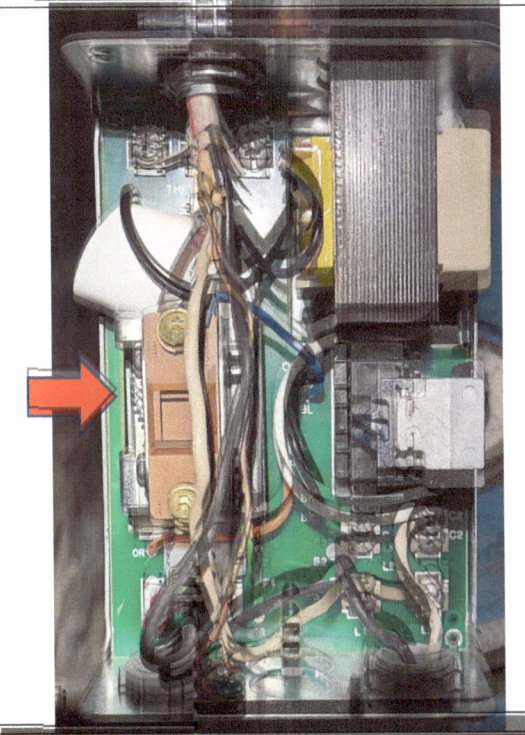

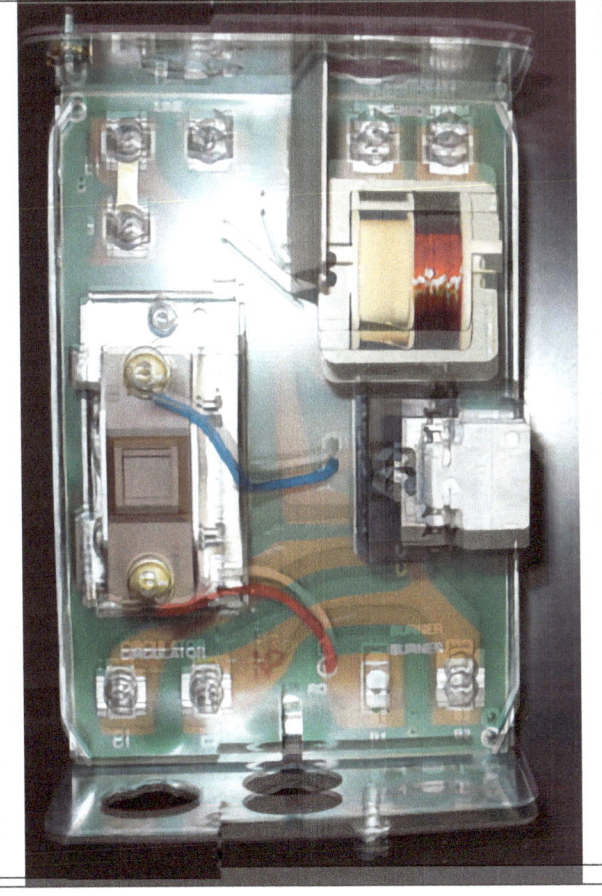

Hydrolevel aquastat temperature controls

Like the Honeywell aquastats, it will control the burner and circulator. In addition, it can be used as a low water cutoff.

Hydrostat cover closed.

Hydronic temperature controls

Most hydronic temperature controls have a sensing bulb inside a well. The well is immersed inside the water. Heat conductive paste is used to better sense the water temperature. The bulb should be touching the well to be in good thermal contact.

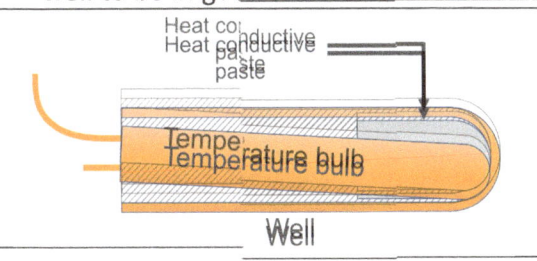

Typical well - They come in various lengths

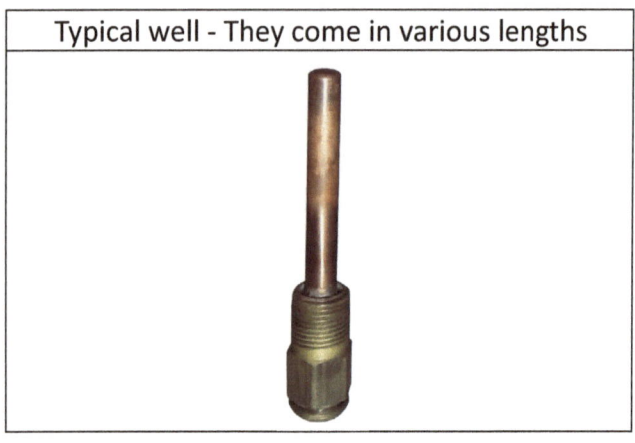

Consider a time delay relay to increase system efficiency

I like installing a time delay relay on the burner. On a call for heat by the thermostat, the circulator & time delay relay TDR are energized. The pump starts and the TDR starts counting down. After the TDR times out, it energizes the burner. I like setting it for 10-15 minutes. In most instances, the call for heat ends without the burner starting. This works on boilers with a higher thermal mass such as a cast iron sectional boiler.

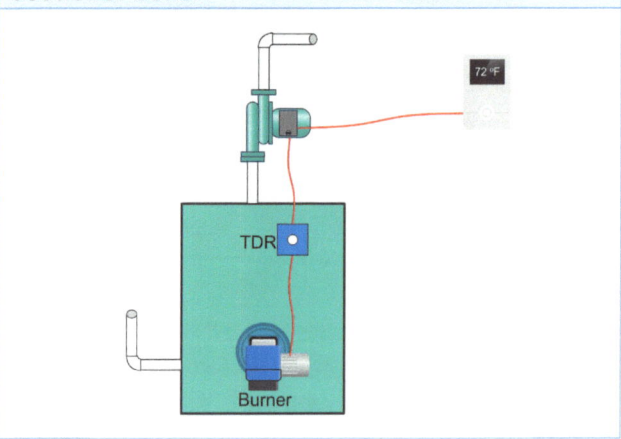

Comparing air to water		
	Air	Water
Specific Heat Btu/lb/°F	0.24	1.00
Density Lb/ft³	0.074	62.4
Heat capacity Btu/ft³/°F	0.018	62.4

The Twenty Degree Rule
Boiler Delta T

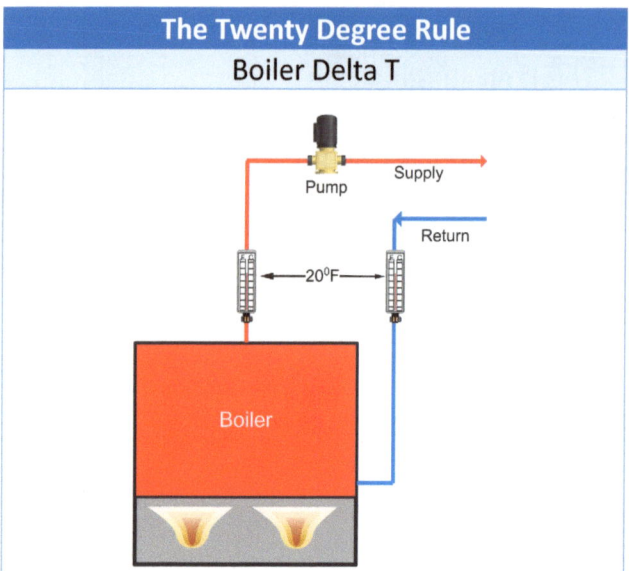

If the boiler Delta T is above 20°F, the water flow may be low, or the burner is not firing at high fire. If the Delta T between is below 20°F, the water flow may be excessive, or the burner is overfiring.

System Delta T

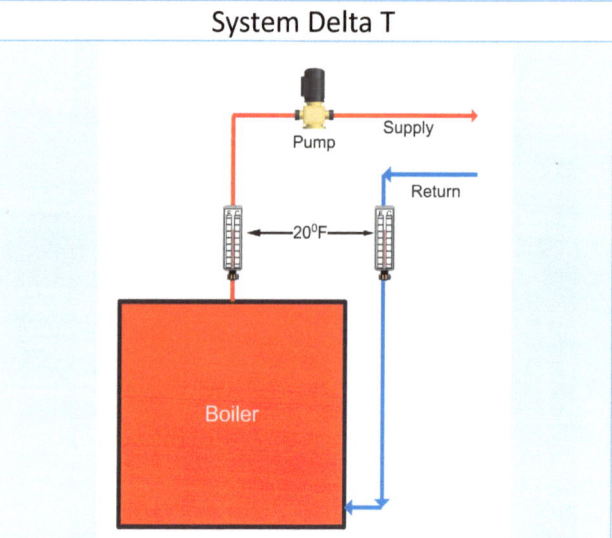

If the system Delta T between the supply and return pipe is above 20°F, the water flow may be low, or the building heat load is high. If the Delta T between is below 20°F, the water flow may be excessive of the building heat load is low.

Hydronic distribution costs are 10 times more efficient than air systems.

Low water cutoffs

Commercial boilers require a low water cutoff which can be tested to verify it shuts off the burner when the water level drops. The test button is required if the area follows ASME CSD1 code. Every boiler should have a low water cutoff. If you come across a residential boiler without one, it should be installed.

Some hydronic boilers don't use a low water cutoff. Below you will see a flow switch used on copper finned boilers. This assures the proper flow through the boiler.

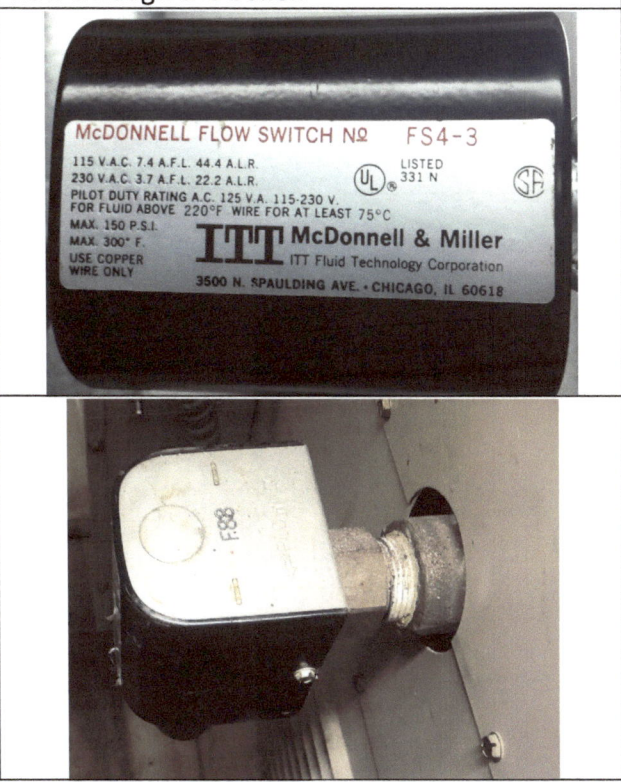

Float type low water cutoffs

Probe low water cutoff

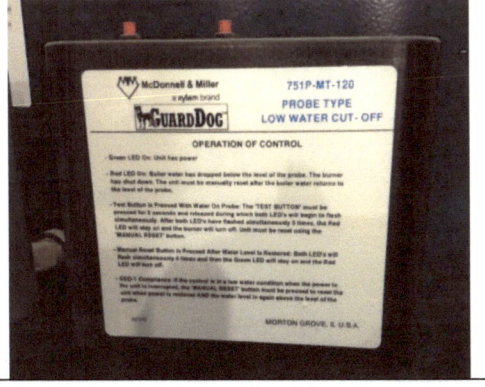

Probe low water cutoff with test and reset buttons

What caused the boiler leak?	
These are the common causes for a leaking boiler.	
Incorrect water treatment	Verify the water treatment system is working and is correct for the boiler metallurgy.
System leak	If the system is leaking water, this allows fresh untreated water into the system which will destroy the boiler.
Thermal shock	If the boiler experiences thermal shock, it can quickly destroy the boiler. Thermal shock is when the Delta T between the return water and the boiler water is too high. It should be less than 20^0F.

Boiler noises

Squealing noise when boiler runs

Defective blower / inducer	The motor bearings may be defective.
Blower wheel	If the blower wheel is misadjusted, it could squeak from rubbing on housing.

Boiler makes a moaning sound when firing

Flexible gas pipe	Sometimes gas flow through the flexible csst tubing is noisy. Verify it is properly sized.
Insufficient water flow	Check flow and temperature rise through the boiler. Temperature rise should be 20-30°F, depending on the boiler.
Scale buildup	Scale formation could decrease flow and water transfer.

The boiler makes a ticking / cracking noise when firing

Scale	Scale could be forming on the water side of the boiler. The water treatment specialist should be consulted.
Low Water	The boiler water level may be too low. If so, do not add water to the boiler until the boiler is shut off and cool.
Sludge inside boiler	Try draining the sludge from the boiler. You may need to hose the sludge from inside the boiler.

The boiler makes a humming or buzzing sound

Bound motor	Look for a bound electrical motor such as the burner blower, induced draft fan or pump.
Loose / broken wire	Check all wiring connections and snug the screws on the terminal. Look for arcing.
Defective relay	Check the condition of the relays. Replace the defective ones.
Wrong voltage	Verify the electrical voltage is within the proper range.
Excess flow	If the flow is too high, it will sound like buzzing or humming.

The boiler makes a hissing sound when firing

Boiler is leaking	Boiler water can be dripping onto burner flame.

The pump is squealing

Pump is defective	Check bearing assembly. Check motor. Dirt inside pump volute

Chattering in pipes

Check valve	Verify check valve is working and not piped backward
Closed valve	Verify all valves are open to the system.

Piping makes a gurgling noise

Air in piping	Air could be trapped in the piping. Bleed air and ensure the air removal fitting works.

> *Installing a water meter on the makeup water pipe to monitor and track makeup water is a good idea.*

Excessive combustion noise

Improper air to fuel adjustment	Verify air to fuel ratio is correct
High gas pressure	Verify correct gas pressure and burner not overfiring.
Excess draft	Check the draft and verify it's not too high.
Fast opening gas valve	If the gas valve opens too quickly, it could cause rumbling. Try partially closing the adjustable orifice on gas valve vent.
Improper flue pipe	Verify the flue pipe is sized properly and has correct pitch.
Blocked flue passage	Verify the flue is free and clear and not blocked.

Main air vent is squealing / noisy

Restricted	Air vent. Check for blockages in air vent outlet
Wrong location	Air vent should have 18 pipe diameters upstream for proper air venting

Pipes are banging

Low water	Verify system pressure is correct and not too low.
Zone valves	If a zone valve closes quickly or causes the pump to dead head, that could cause banging.

Pump is grinding / banging

Impeller	Check impeller to see if its defective or volute is dirty
Oil needed	Add oil to pump motor or bearing assembly
Defective pump coupling	Replace pump coupling
Defective bearing assembly	Replace bearing assembly or pump

23

Excessive vibration in system	
Defective burner / pump motor	Check the motor bearings. Replace the motor if defective.
Improper combustion	Check the air to fuel ratio
Closed valve	Verify the pump discharge pipe is not plugged or restricted. Look for closed valves.
Blower wheel	If the blower wheel is dirty or some of the balancing weights are missing, it could cause vibration. Replace blower wheel before it ruins blower motor bearings.

Baseboard radiation is ticking	
Slides missing from under the element	Look for slides (1) under the heating element that allow it to expand and contract
Hole for piping is incorrect	The tubing may be too close to the end of the hole for the vertical piping (2). The tubing will tick as it expands and touches the edge of the hole.

Zone valve is banging	
Excess flow	Valve trying to close against excess pump flow. Try slowing down pump or installing a bypass pipe.

Common Hydronic Piping

What is Pumping Away?

Pumping Away means to pump away from the expansion tank toward the system. This piping method:
- Reduces air in system.
- Reduces system noise.
- Increased efficiency
- Reduced corrosion in the system

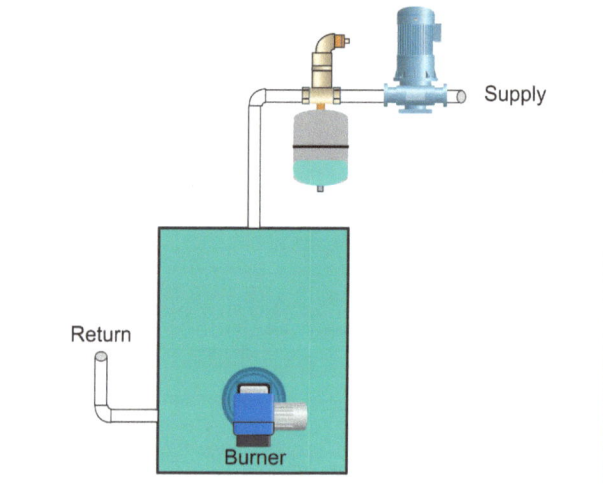

Pumping toward the expansion tank causes the following:
- Increases air in system.
- Noisy operation
- More internal corrosion
- Reduced system efficiency
- Having to bleed radiators more often

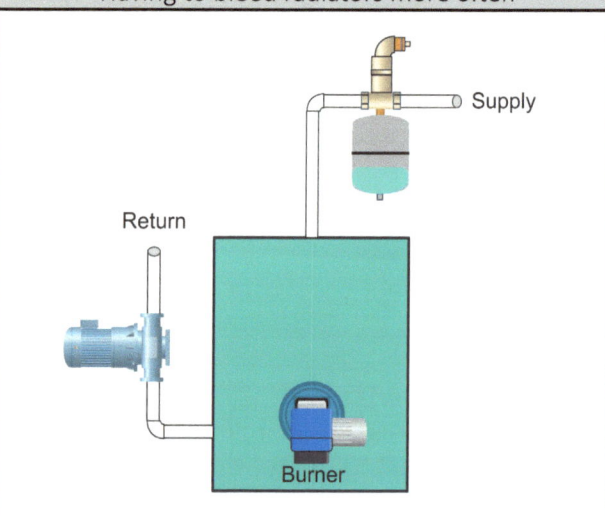

Water can hold almost 3,500 times more heat than air for the same volume.

Why have the circulator on the supply?

Circulator on the return

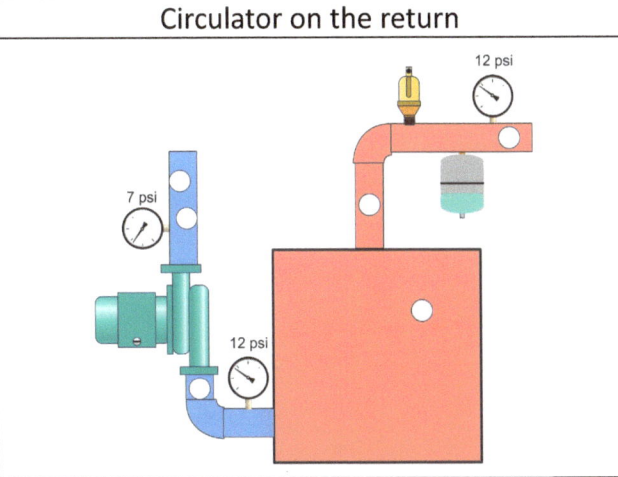

Think of a hydronic system like a finely manicured lawn. Just like weeds, the air bubbles are always there but under control. The system static pressure, 12 psi, keeps the air bubbles small and manageable. I'm using 12 psi static pressure as it's the most common static pressure for a two-story building. This allows the bubbles to be contained and removed by the air removal fitting and vent.

For water to flow, the circulator creates a pressure differential. In this case, the pressure differential is 5 psi. If the circulator is on the return, the pump outlet is facing the expansion tank. The extra 5 psi generated by the circulator is absorbed by the expansion tank. Undaunted, the circulator develops the pressure differential by lowering the inlet pressure. The upstream pressure of the circulator drops to 7 psi and that's when everything falls apart. The bubbles that were in submission now expand, become more buoyant, and can't be removed by the vents. Essentially, you lawn is filled with weeds. Those air bubbles are free to munch on the piping and boiler and restrict the water from getting where it needs to go. The owner or service tech has to vent the radiators more often.

Circulator on the supply

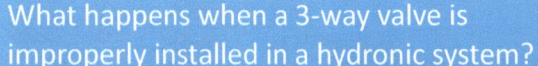

When we have the circulator on the supply, *pumping away* from the expansion tank, the outlet pressure increases by 5 psi. This makes the circulator discharge pressure 17 psi, and the inlet stays at 12 psi. By increasing the system pressure, the bubbles shrink even more, making them easier to remove. This results in even less air venting required by the owner or service tech.

What happens when a 3-way valve is improperly installed in a hydronic system?

The boiler was shocked, and the cast iron sections destroyed.

A well-designed hydronic system with a high efficiency circulator can deliver heat using less than 10% of a forced air system using a blower.

Direct return

In a direct return system, the first radiator fed is the first to return. This causes that area to overheat and areas further away to underheat. Flow has to be controlled using balancing valves.

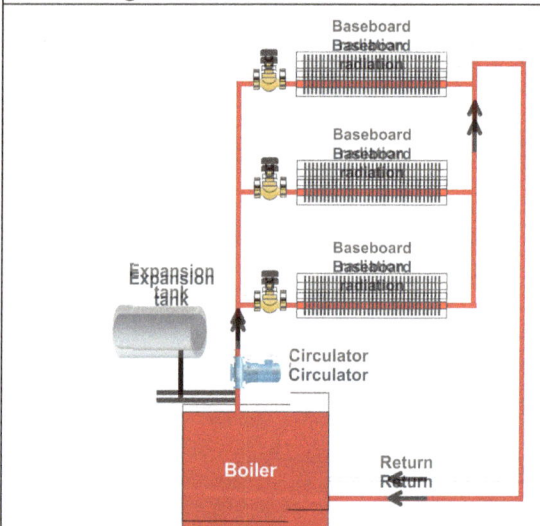

Advantages: Lower installed cost
Disadvantages: Comfort complaints, difficult to balance.

Reverse Return

In these systems, the first radiator fed is the last returned which makes it almost self-balancing.

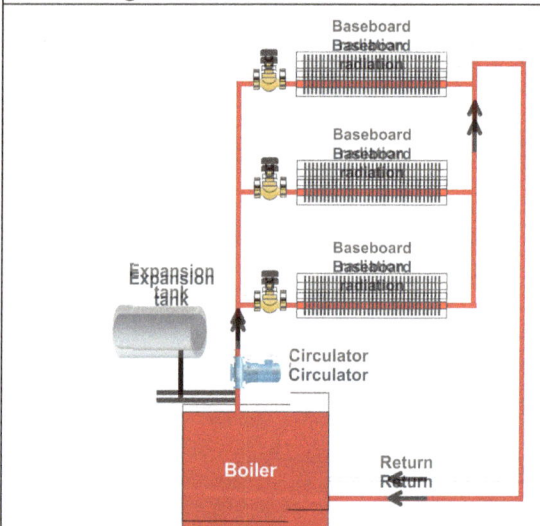

Advantages: Almost self balancing, even temperatures across the system.
Disavantages: Higher installation cost

Ghost Flow

Ghost flow occurs when water goes backward through an idle zone when another is calling for heat. If you see below, Zone 1 has a call for heat and water is flowing backward through Zone 2, overheating the space. A weighted check valve will often resolve this.

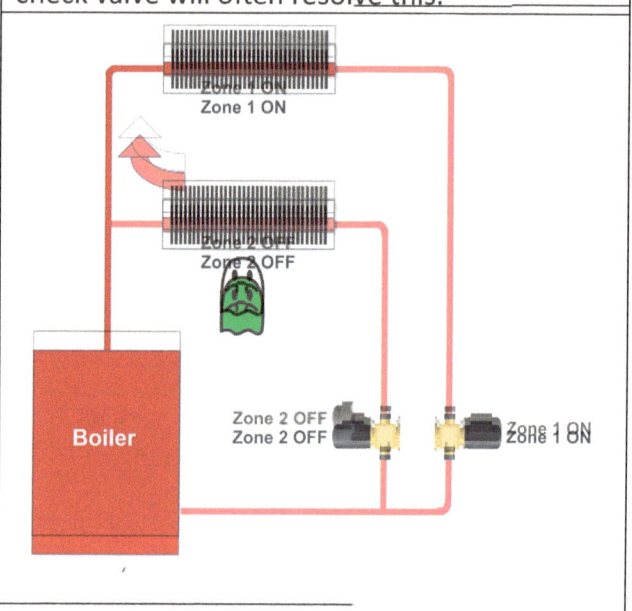

Weighted check valve:

> An adult in light activity will generate about 400 Btu/hr.

Pipe GPM @ Various Velocities Feet per Second

Pipe Size Inches	2 FPS	3 FPS	4 FPS	5 FPS	6 FPS
	Gallons per Minute				
1/2"	2	4	5	12	22
3/4"	4	6	9	21	38
1"	7	10	14	35	62
1 1/4"	12	18	24	60	108
1 1/2"	16	24	33	81	147
2"	27	40	54	135	242
3"	59	89	118	296	533
4"	102	153	204	510	918
6"	231	347	463	1,157	2,082
8"	400	600	800	2,000	3,599
10"	631	946	1,261	3,153	5,675
12"	895	1,343	1,791	4,476	8,058
14"	1,083	1,624	2,165	5,413	9,744
16"	1,413	2,120	2,825	7,065	12,717
18"	1,789	2,684	3,579	8,947	16,104
20"	2,222	3,333	4,444	11,110	19,998

Recommended velocity for hydronic system piping 2-6 feet per second.

30°F Delta T = Btuh output of boiler /15,000 = GPM

20°F Delta T = Btuh output of boiler /10,000 = GPM

Primary – Secondary

Most boiler manufacturers suggest piping the boiler in a primary-secondary fashion. This allows the isolation of the idle boiler and a reduction in energy loss.

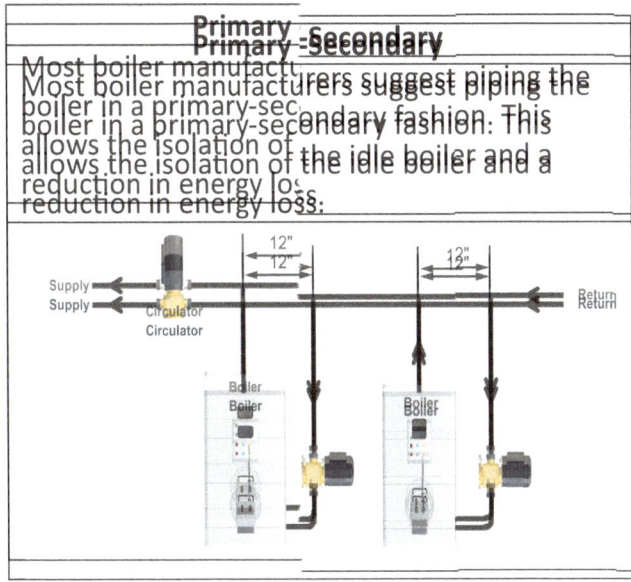

Flow Through Design

Older hydronic systems were piped using a flow-through design. There are more efficient ways to pipe hydronic systems. The drawback to this is the idle or lag boiler has flow through it all the time increasing the heat loss to the boiler room and chimney.

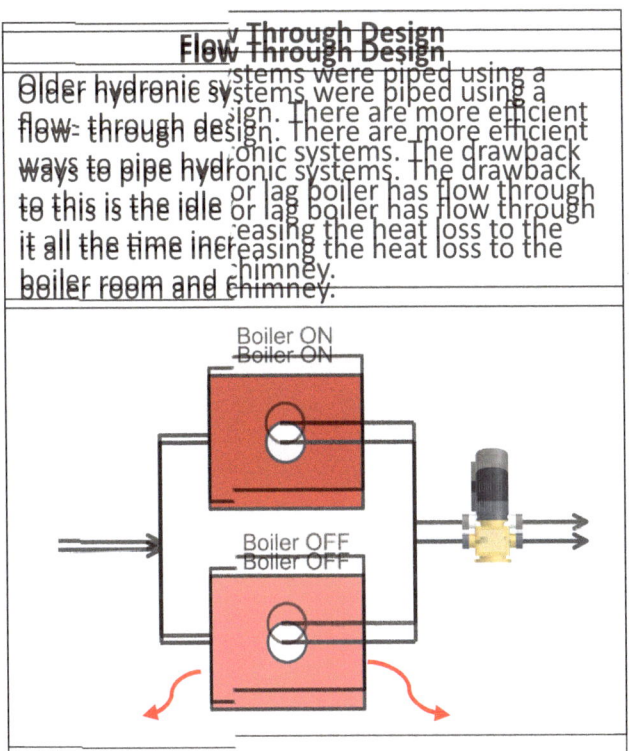

Manifolded Primary Secondary

An even more efficient way to pipe multiple hydronic boilers is using a manifolded primary secondary piping. The advantage to this each boiler gets the same temperature and the system has higher efficiency.

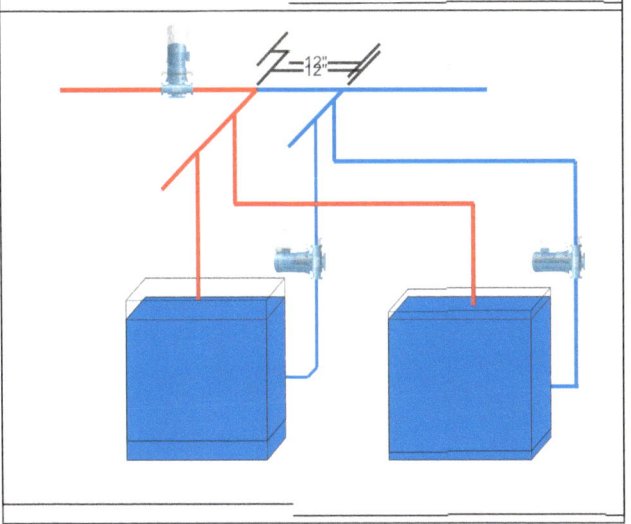

Heat travels from hot to cool.

Boiler piping to help purge air

By piping the boiler like this, it allows you to quickly purge the air from the system. To purge the air, do the following:

Shutoff circulators
Connect hose to purge valve and direct the other end of the hose to the floor drain.
Close service valve, 1
Close isolation valves 2, 3
Open makeup water valve
Open purge valve
Open isolation valve to one zone, 2
Run water until the air is gone.
Close valve 2 and open valve 3.
Run water until the air is gone.
Close purge valve
Open valve 1, 2, and 3 and start system.

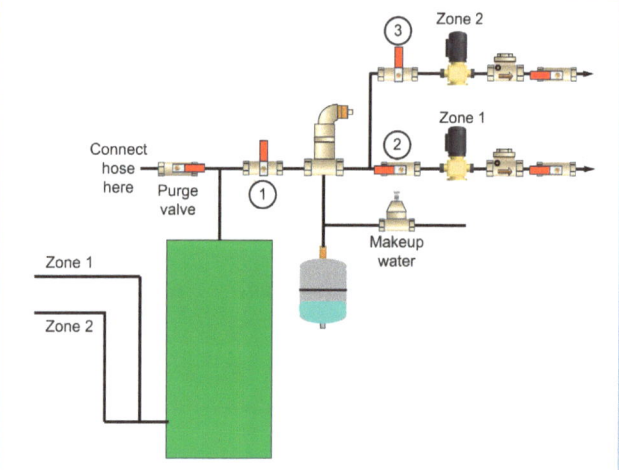

This piping will allow purging of the system air in the boiler room. This will save you time venting.

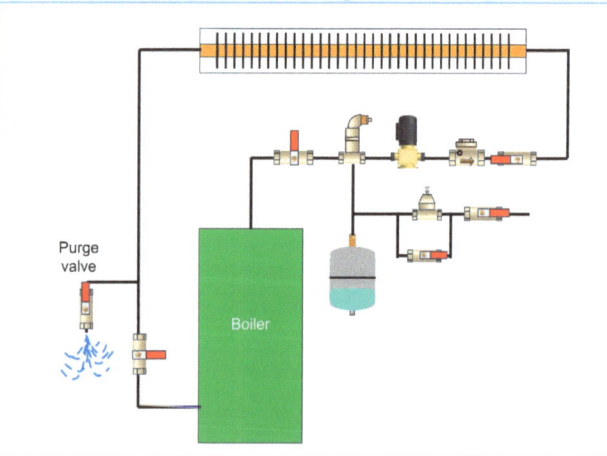

Estimated HVAC Equipment Life

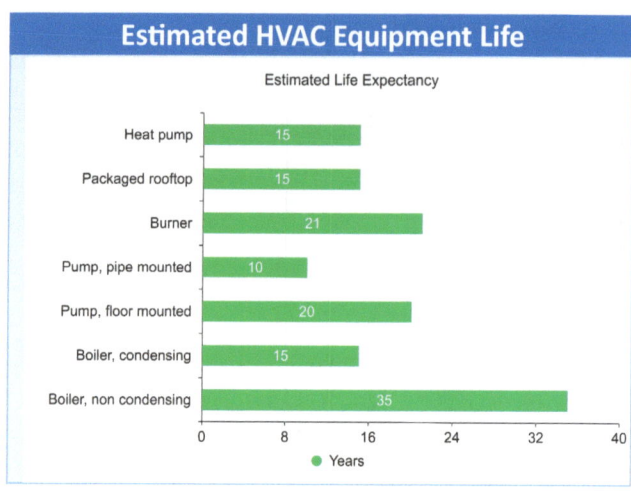

Filling a system without a bypass valve

If there is no bypass on the water feeder, you may be able to feed the water quickly by connected a washing machine hose and garden hose to the boiler drain valve.

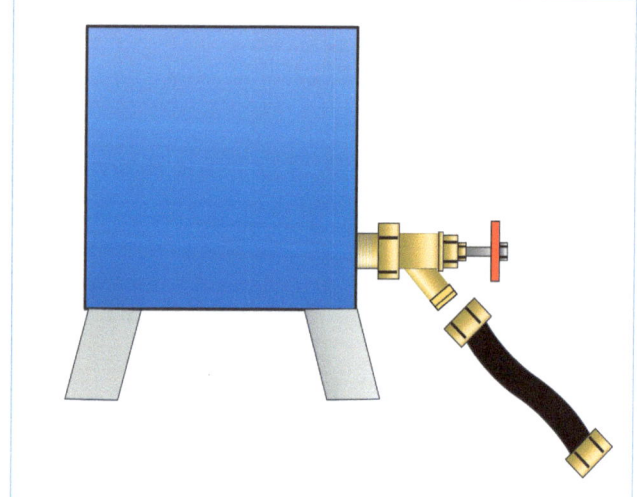

A ¾ inch tube carrying water can transport as much heat as a 14" x 8" duct.

Troubleshooting hydronic piping

Verify the piping is large enough to handle the Btu's. 2-6 feet per second is the recommended velocities. Velocities greater than 6 FPS could causes erosion of the pipes interior & noisy operation.

Is there a 3-way valve? A 3-way valve is used to temper the supply water temperature by mixing the return water with the supply water from the boiler. This valve could severely damage the boiler by either thermal shock or insufficient flow through the boiler if incorrectly controlled. To avoid shocking and damaging the boiler, some manufacturers suggest installing a blend pump which will inject hot supply water into the return pipe to warm the return water.

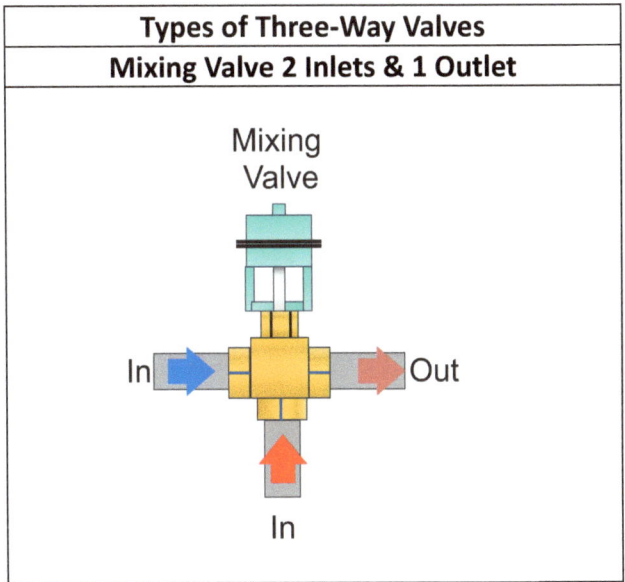

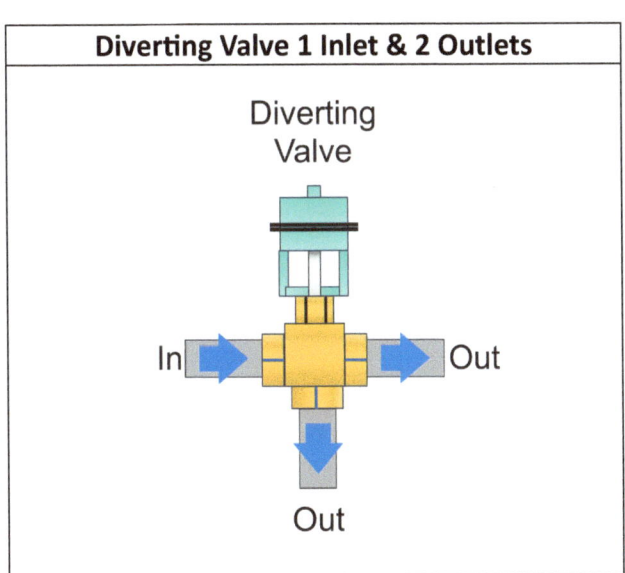

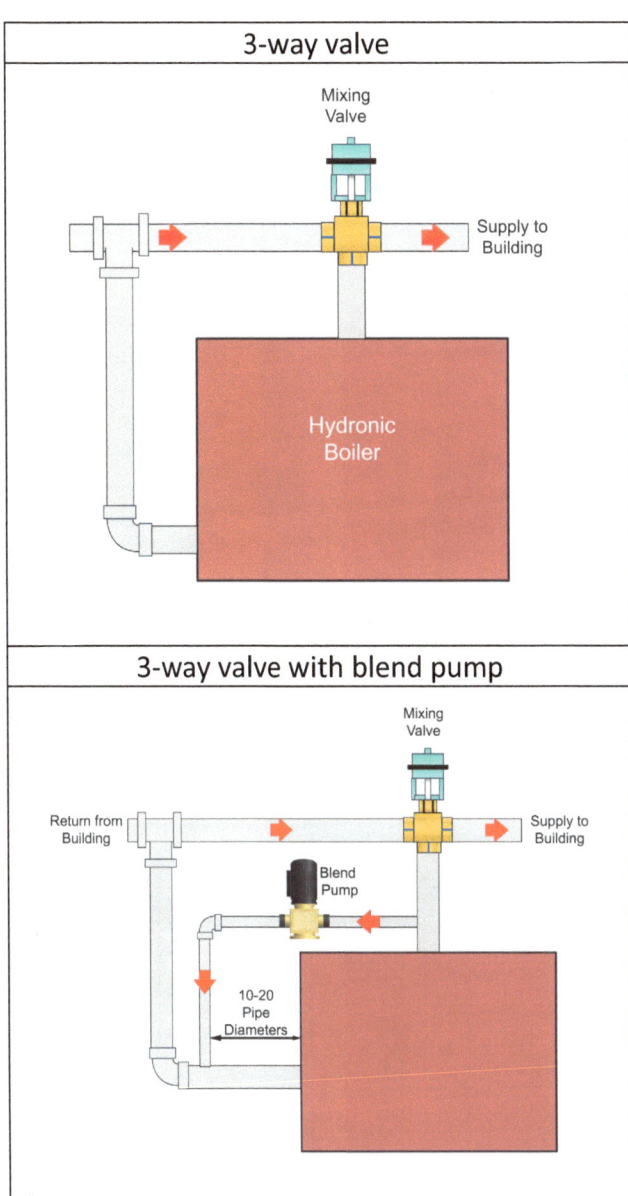

Sizing a bypass pump when using a 3-way valve on a cast iron boiler
¼ to 1/3 the GPM of the system circulator
Bypass pump termination should be 10-20 pipe diameters upstream of boiler return connection
For example: A 500,000 Btuh boiler has a 50 GPM pump based on a 200F Temperature Rise. The bypass pump GPM would be between 12 ½ and 16 ½ GPM.

For half the heating season, the system has a heat load of 1/3 or lower.

Buffer Tank A buffer tank is sometimes installed when using low mass boilers. Low mass means the boiler water content is less than a traditional boiler. The buffer tank reduces the boiler cycling, increasing the longevity and efficiency of the boiler.

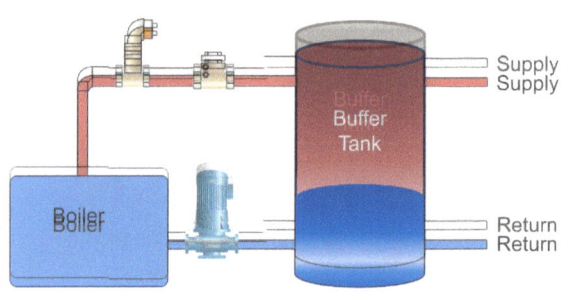

Buffer Tank Sizing
Based on 10°F temperature difference in tank

Btuh Difference	Minimum boiler run time in minutes		
	10	15	20
	Minimum Gallon Tank Capacity		
2,500	5	7.5	10
5,000	10	15	20
7,500	15	23	30
10,000	20	30	40
12,500	25	38	50
15,000	30	45	60
17,500	35	53	70
20,000	40	60	80

Sizing a buffer tank
The formula for sizing a buffer tank is:

$$Gal = \frac{Time \times (Min\ boiler\ output - Smallest\ load\ Btuh)}{Temperature\ difference\ in\ tank \times 500}$$

When sizing a buffer tank, you will need to know the following:

- Minimum run time of boiler, typically 10-20 minutes. The longer the run time, the higher the system efficiency.
- Minimum boiler output
- Minimum system load
- Temperature difference in tank, typically 10-15°F

Let's look at sizing a buffer tank for the following system:

Boiler: 80% efficient boiler 125,000 maximum input with a 5 to 1 turn down burner. At low fire, the boiler will be at 25,000 Btuh input and 20,000 Btuh output.

The smallest zone is 5,000 Btuh.

10 minutes X (20,000 – 5,000) = 150,000

150,000 Div. by (10 deg x 500) = 30-gallon tank

Mixing Valves
In some applications, you will need a mixing valve to raise the return water temperature above the condensing temperature, typically 140°F.

% of Supply water needed to raise return water to 140°F

HWS °F	Return water temperature			
	135°F	125°F	115°F	105°F
180°F	12.5%	37.5%	62.5%	87.5%
170°F	16.7%	50%	83.3%	NA
160°F	25%	75%	NA	NA
150°F	50%	NA	NA	NA

Determine % of Hot Water needed to raise water temperature.

$$\frac{MWT - C}{HWT - C} = \frac{140 - 50}{180 - 50} = 90/130 = 69.2\%$$

C = Cold water temperature
MWT = Mixed Water Temperature
HWT = Hot Water Temperature

> Most pumps require 5-10 pipe diameters of straight pipe upstream of pump suction for best operation.

Connecting new boilers to old pipes

When connecting new condensing boilers to old hydronic systems, the system was probably designed for 180°F. This meant the system was designed to heat on the coldest days using 180°F water. If you connect a condensing boiler to the loop, it may not be in the condensing mode for much of the heating season. If you look at the bin temperatures for Pittsburgh Pa where I live, the boiler doesn't start condensing until the outdoor temperature reaches about 49°F.

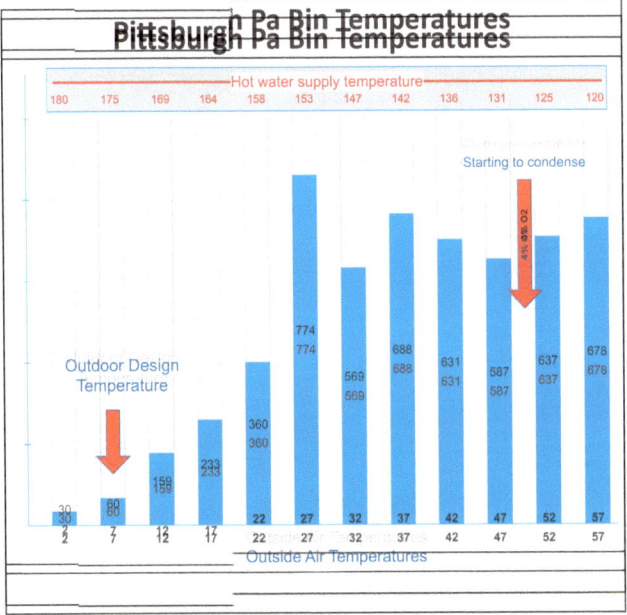

Pittsburgh Pa Bin Temperatures

Trouble-shooting Hydronic Systems

Troubleshooting hydronic systems can be challenging as the pipes are hidden behind walls. Your imagination has to help you "see" the routing of the piping. In addition, a component in one part of the system could affect the operation in another. An example of this could be if the relief valve is leaking. A faulty relief valve may cause this, or it could be caused by a flooded expansion tank or a malfunctioning pressure reducing valve. You would hate to replace the relief valve and have the same issue.

Tools every hydronic tech should have:

Kneepads - Trust me, the older you will thank you.	
Radiator Key - Most radiators require this to bleed the air.	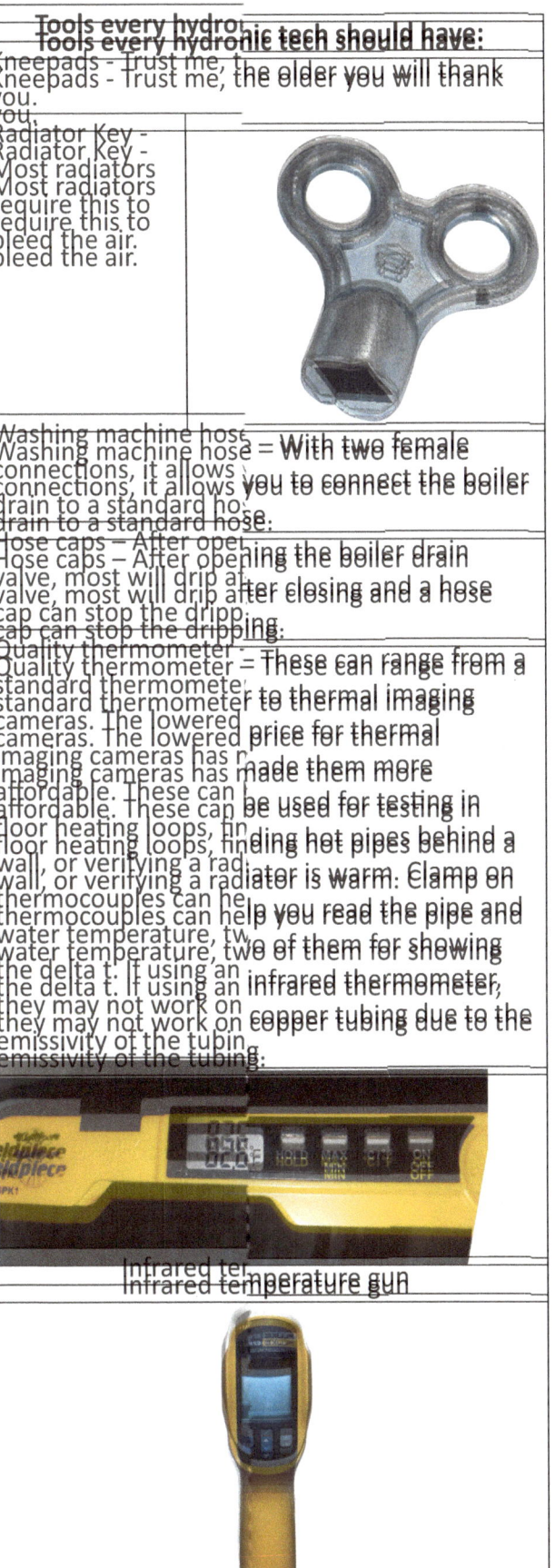
Washing machine hose – With two female connections, it allows you to connect the boiler drain to a standard hose.	
Hose caps – After opening the boiler drain valve, most will drip after closing and a hose cap can stop the dripping.	
Quality thermometer – These can range from a standard thermometer to thermal imaging cameras. The lowered price for thermal imaging cameras has made them more affordable. These can be used for testing in floor heating loops, finding hot pipes behind a wall, or verifying a radiator is warm. Clamp on thermocouples can help you read the pipe and water temperature, two of them for showing the delta t. If using an infrared thermometer, they may not work on copper tubing due to the emissivity of the tubing.	

Infrared temperature gun

Thermal imaging camera

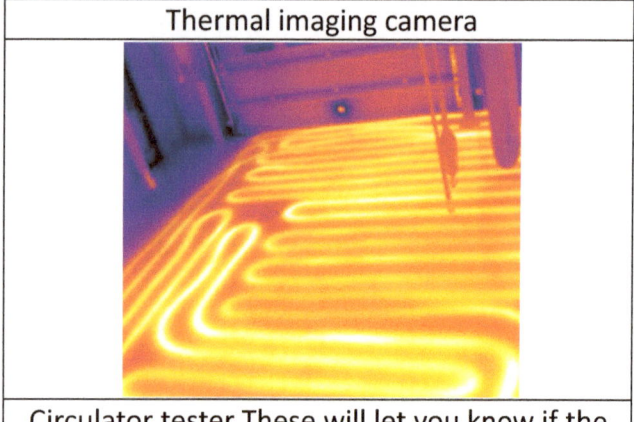

Circulator tester These will let you know if the pump or solenoid has power to it.

Free app from Danfoss. This app works lets you know if there is power to a circulator or solenoid valve.

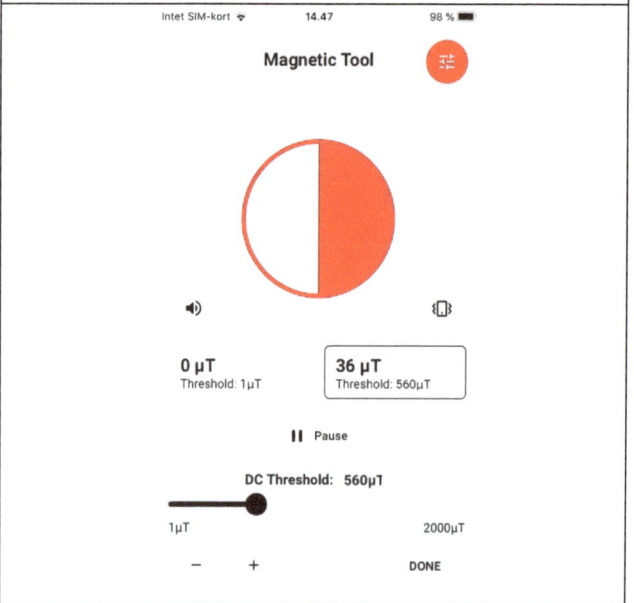

Water expands 3% of its volume when heated from room temperature to $180^0 F$.

Do the next tech a favor, install isolation valves so the entire system doesn't have to be drained when working on the boiler and the nearby components.

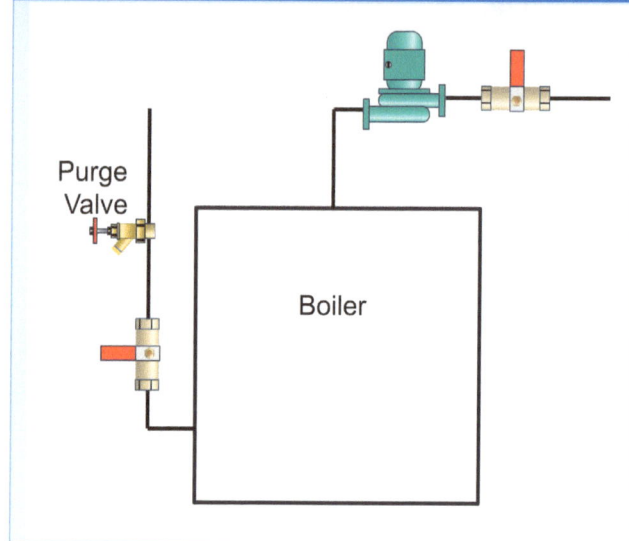

System pressure reducing valve with bypass

This is worth the extra time & materials to install a bypass.

This is not a Friday afternoon project

Looking around the system

Ray's Rule #1 Always assume the old boiler is installed incorrectly. This has helped me in many projects.

Are the radiators enclosed? Some radiator enclosures will affect the heat output of the radiator.

Is there a water treatment system? Every boiler needs water treatment.

Is there a water meter on the makeup water? I like using a water meter on the makeup water pipe. It will help the owner know if there is a system leak.

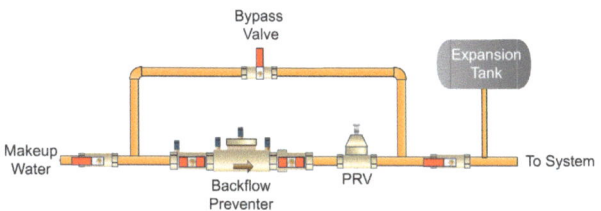

Bypass valve A bypass valve around the system pressure reducing valve PRV is a good idea as it makes filling the system much quicker.

Hydronic replacement boilers are sized for the heat loss of the building. To verify the calculations, I check the supply and return pipes sizes and the system pump ratings. Ever since December 1899, most hydronic systems in the US were designed using 180^0F as the design temperature which means the boiler should operate at that temperature on the coldest days.

> *The early hydronic systems didn't use a pump to distribute they heat. They relied on the buoyancy of the water. Hot water rose, cold water dropped.*

Using the PTA gauge to troubleshoot

The PTA or Tridicator gauge could help diagnose system problems.

Pressure - These are the numbers next to the Red Arrow. It shows the static pressure inside the system. The static pressure is how high the water is in the system. It requires one pound of pressure to raise water 2.3 feet. As a rule of thumb, I like dividing the height of the highest radiator by 2 and that's the required static pressure. On most residential two-story homes, the static pressure is typically 12 psi.

This gauge also has an altitude reading (Blue Arrow). It shows the height of the water in the building. At 12 psi, the water should be about 27 feet above the boiler.

The pressure reading (Red Arrow) can help you diagnose if the expansion tank is full. If the pressure rises as the burner fires, it indicates the expansion tank is flooded or undersized.

The green arrow shows the temperature of water inside the boiler. On non-condensing boilers, the water temperature should be above 140^0F when the boiler is firing.

Hydronic Zoning

One advantage to using hydronic systems is the ability to zone different areas of the building. The most common ways to do the zoning is either using zone valves or zone circulators. There are pros and cons of each system and staunch supporters of each method. It would take an entire book to describe each system and I will not get into it here. Rather, I am showing some of the drawbacks of each system and how it caused system issues.

Zoning with Pumps

Zone systems using circulators are reliable and provide some redundancy as if one circulator stops working, you still have heat in the other zones. The main drawback to using circulators for zoning is the energy consumption of the circulator motors.

The following system was installed incorrectly and only lasted a few years. When only one zone was calling for heat, there was not enough flow through the boilers. This caused thermal shock and damaged the boiler.

A primary secondary piping arrangement like this could ensure proper flow through the boiler and a longer life for the boiler. Always consult the IOM manual.

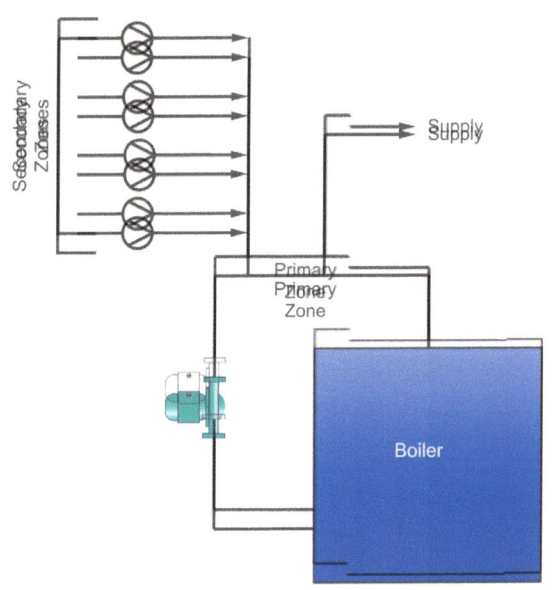

Circulator parts

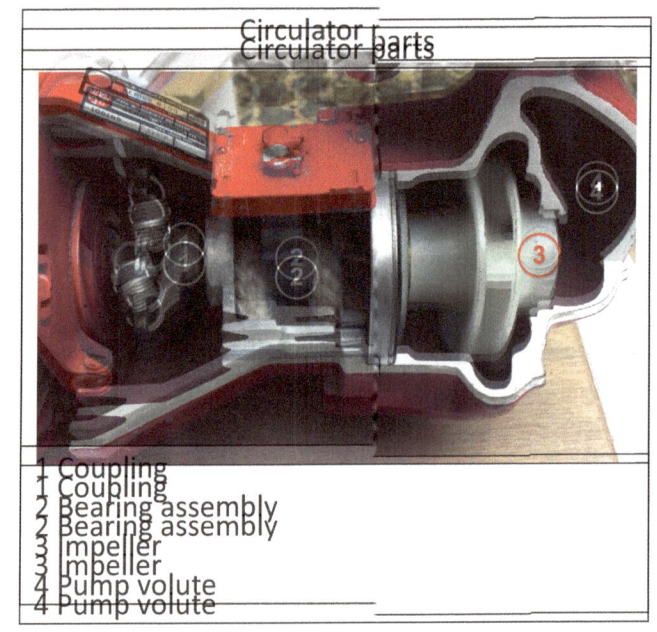

1 Coupling
2 Bearing assembly
3 Impeller
4 Pump volute

When installing hydronic zone valves, I like choosing ones with a manual bypass which allows you to manually open the valve in the event of a malfunction.

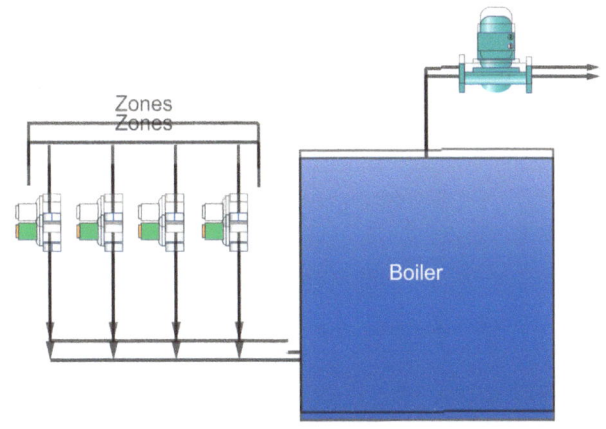

Zoning with Valves

Zoning with valves is a reliable system with lower energy consumption and requires less space and maintenance than circulators. They can be a bit tricky as the flow rate of water through the pipes changes when each zone valve opens or closes. When a pump or circulator pumps against a closed valve, it's called Dead Heading, and it could damage the pump. Dead heading causes increased pressure and temperature inside the pump. Insufficient flow through a boiler can damage it.

On this project, the velocity was too high when only one zone called, and the space never warmed up. To get the apartment to heat, I had to partially close the manual valves to reduce velocity.

To avoid dead heading the pump when using zone valves, there are a couple of options:

Variable speed circulator – Circulators are available that will automatically adjust the speed according to the pressure or delta T in the piping.

Bypass pipe – Some buildings use a bypass pump to ensure there is flow even if the zone valves are closed. The bypass could be manual or automatic. The automatic bypass uses a pressure sensor to modulate the valve opening. A three-way valve is sometimes used instead of two-way valves to eliminate dead heading the circulator.

Manual bypass pipe and valve

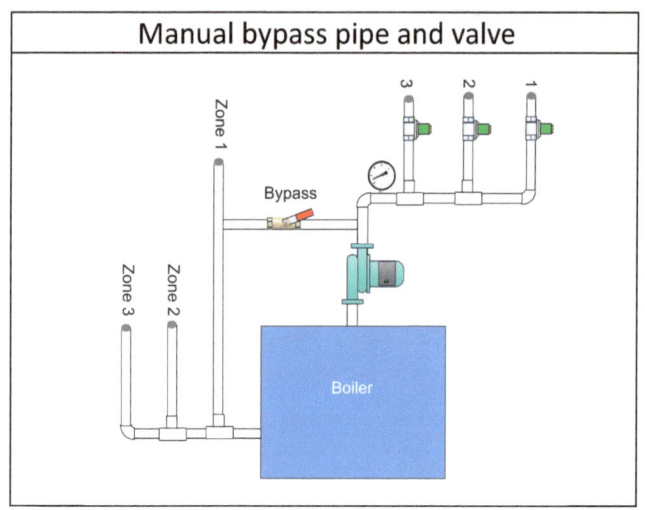

Using a three way valve to bypass

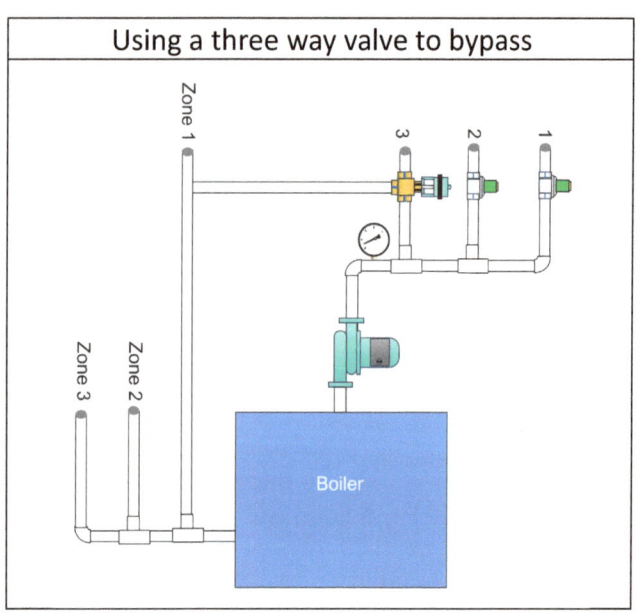

Automatic bypass

Hydronic balancing valves

Air Venting Tips I Learned

Try venting with the pump off. It allows the natural buoyancy of the air to find the highest places.
Verify the system pressure is high enough to reach the top radiators. A quick rule of thumb is to divide the height of the highest radiator by and that will be the correct static pressure.
Start with a higher pressure than required because the system pressure will drop as you vent the air.
There is much less air in systems which pump away from the expansion tank.
I like to use a ¾" ball valve on top of the riser to quickly remove the air.
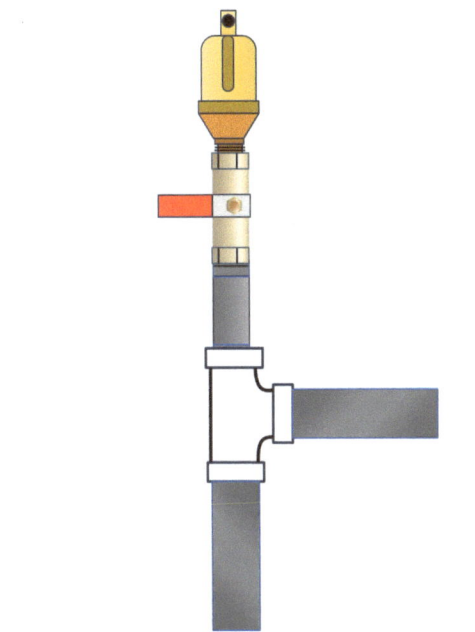
Remember to take a radiator key, screwdriver, and an old cup along with you. Some vents require the key while others use a flat head screw.
If you have a stubborn air pocket blocking the water, try cycling the pump on and off or raising the system pressure.
Be sure the system has an air removal fitting in the piping. This will eliminate the air and quiet the operation.
Every automatic vent leaks after time. I like having a ball valve between the vent and the piping to allow isolation in case you need to replace the vent.
Use a hose to fill the system, much faster than filling through the feeder. Washing machine hoses work great for this.
If you are not getting air, it's not an air problem - Dan Holohan
Microbubble air vents work really well.
Consider installing a purge valve in the boiler room or basement.
Look for hidden air traps like the one shown in the drawing below. This could trap air and almost impossible to vent. The solution was to install an air vent in the high piping.
Fan coil without vent
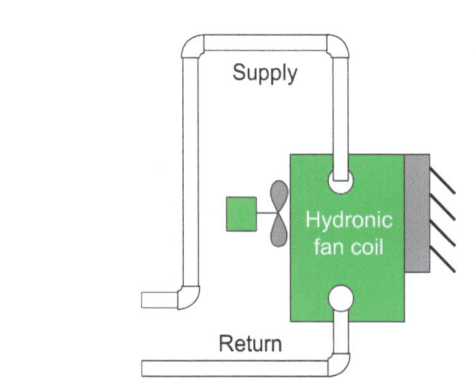
Fan coil with air vent
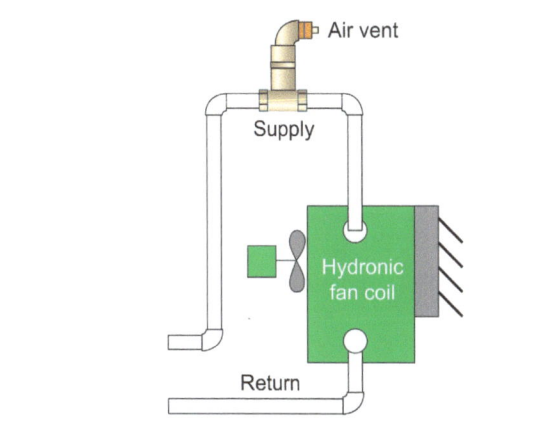

Dissolved oxygen is 10 times more corrosive than CO_2.

Understanding condensing boilers

When troubleshooting condensing hydronic boilers, there are some differences between condensing and non-condensing boilers. The following are some of the things I have learned.

Sometimes, condensing boilers don't condense:

A condensing furnace will always condense because of the cooler air temperature passing over the secondary heat exchanger. That doesn't happen with a hydronic system. The hydronic system was most likely designed for a 180°F design temperature. This means the heating distribution system was designed to heat the building on the coldest days using 180°F water. The second factor is the boiler will not start to condense until water temperature drops to around 130°F. This is called the dewpoint temperature. The condensing boiler will not reach 90% efficiency until the water temperature is about 80°F.

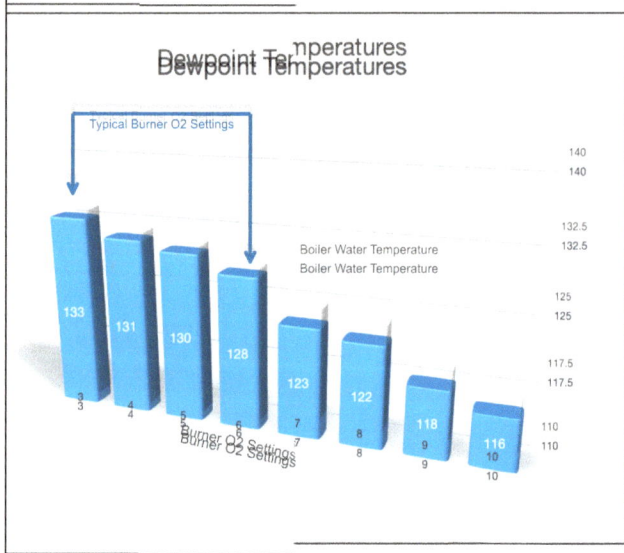

Condensing boilers don't last as long as the old non-condensing ones:

The life expectancy of a condensing boiler is about half that of a non condensing boiler. This should be explained to the owner.

Condensing boilers require more maintenance than non-condensing ones:

Old cast iron boilers with a standing pilot would run for years with little or no maintenance. The condensing boilers require much more maintenance than those old units. When talking with the owner about changing the boiler to a condensing one, the additional maintenance should be discussed. The condensing boiler can fail quickly without proper maintenance and the owner will blame the installer. The condensing boiler should be opened, and the heat exchanger and burner cleaned every summer.

Condensing boilers are temperamental than the old type boilers.

The condensing boiler is more prone to lockouts from variations in water flow and electrical voltage than the old boilers with mechanical controls. Be sure the electrical supply is stable.

The tolerances inside the boiler and circulators are much tighter than the older type units. High efficiency strainers should be installed when installing the boiler.

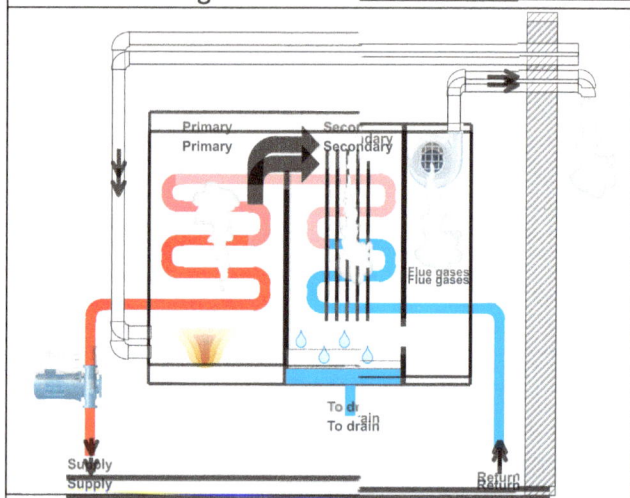

Air vents should have 18 pipe diameters upstream of the vent to properly work.

Understanding the Boiler Sequence of Operation

Atmospheric burner

- The electrical power is switched on and there is a call for heat.
- Presence of water is verified.
- The pilot flame is ignited and verified. Most atmospheric burners use a flame rod placed inside the pilot flame to verify the pilot flame lights.
- Once the pilot is verified by the flame safeguard, a signal is sent to open the gas valve(s).
- The burner will stay lit until either the call for heat ends, the water temperature rises high enough to meet the setting of the operating temperature control, or one of the safety controls opens.

Power burner On Off

- The electrical power is switched on and there is a call for heat.
- Presence of water is verified.
- The burner blower motor starts & verified, typically using a differential pressure switch.
- Once the blower is verified, the burner runs 30-90 seconds to purge any leftover fuel inside the boiler or chimney. This is called Pre-Purge.
- At the end of the pre-purge, the ignition transformer is energized, and the ignition electrodes starts to spark. The ignition transformer is rated for 6,000 to 12,000 volts. At the same time, the pilot solenoid valve opens, and the spark ignites the pilot flame. The pilot flame is verified using a flame rod, infrared, or ultraviolet sensor.
- After the pilot operation is verified, the main gas valve(s) open.
- The burner will stay lit until either the call for heat ends, the water temperature rises high enough to meet the setting of the operating temperature control, or one of the safety controls opens.

Typical Sequence for Boilers with Variable Stage Power Burners

Low High Off Power Burners After the pre-purge, the burner will travel to low fire & ignite & verify the pilot. When the pilot is verified, the flame safeguard will open the electric shutoff valves. Once the flame is established, the burner drives to high fire and will stay there until the call for heat ends or the boiler water temperature meets the control setpoint.

Typical Sequence for Low High Low Power Burners After the pre-purge, the burner will drive to low fire, ignite & verify the pilot flame. When the pilot is verified, the flame safeguard will open the electric shutoff valves. After the flame is established, the burner drives to high fire to meet the setpoint of the firing rate control. Once the water temperature reaches the setpoint of the firing rate control, it will then drop to low fire. The burner will travel between low and high fire as directed by the firing rate control until the call for heat has ended or the boiler temperature reaches the pressure control setpoint.

Typical Sequence for Modulating Power Burners After the pre purge, the burner drops to the low fire position, ignites & verifies the pilot flame. When the pilot is verified, the flame safeguard will open the electric shutoff valves. After the flame is established, the burner drives to high fire to meet the setpoint of the modulating control. When the water temperature approaches the modulating control setpoint, the control will position the burner between low and high fire to maintain the desired water temperature. It will continue adjusting the firing rate until the call for heat ends or the water temperature meets the operating control setpoint.

> The leading cause of boiler accidents is improper maintenance.

Boiler Sizing

Heating Water with Electric

The owner would like to size an electric boiler for the following conditions:
Raise water temperature from $140°F$ to $185°F$
Water flow is 10 GPM.

$$KW = \frac{GPM \times 500 \times \Delta t}{3413}$$

500 = (1gpm x 8.345 lbs./gal x 60 minutes)
GPM = Flow of water in gal/min
Δt = Temperature rise
3413 = Btuh to KW

$$KW = \frac{10 \times 500 \times 45}{3413}$$

$$KW = \frac{225,000}{3413}$$

KW = 65.9

65.9 x 1.2 Safety Factor = 78.96 KW

Sizing Heater for Snow Melt
According to EngineeredToolBox.com

	Concrete	Asphalt or Blacktop
Entering water °F	120-130°F	120-130°F
Distance between coils	No more than 12"	No more than 9"
Thickness above tubes	1 ¼" to 1 ½"	1 ½"
Maximum Length of coil	½" tube – 140 feet	½" tube – 140 feet
	¾" tube -280 feet	¾" tube -280 feet

100 Btu per hour per square foot would melt up to 1 ½" per hour snowfall

Capacity is 200 Btuh/foot if ½" tube length is no more than 60 ft and ¾" tube no longer than 150 ft.

Sizing Heater for Hot Tub

Up to 10 feet square and 4 feet deep – 2,000 to 2,500 Btuh with cover on

6,000 to 9,000 Btuh loss per hour when cover is off.

Sizing a replacement hydronic boiler

Tips for sizing a replacement hydronic boiler

Assume the existing boiler installation is wrong.

Boiler should be sized for the building's heat loss.

Verify the piping and the pump(s) you are connecting to are sized to handle the Btu heat loss.

Try keeping the water velocity around 4 FPS.

Is there an air removal fitting?

If reusing the existing expansion tank, verify its sized properly. If the existing one is flooded, find out why. Does the tank have a hole, gauge glass washers dried out and leaking?

Convert Flow (GPM) to Velocity (FPS)

GPM	Pipe Diameter in Inches					
	2	4	6	8	10	12
	FEET PER SECOND					
5	0.51	0.13	0.06	0.03	0.02	0.01
10	1.02	0.26	0.11	0.06	0.04	0.03
15	1.53	0.38	0.17	0.10	0.06	0.04
20	2.04	0.51	0.23	0.13	0.08	0.06
30	3.06	0.77	0.34	0.19	0.12	0.09
40	4.08	1.02	0.45	0.26	0.16	0.11
50	5.10	1.28	0.57	0.32	0.20	0.14
60	6.12	1.53	0.68	0.38	0.24	0.17
70	7.14	1.79	0.79	0.45	0.29	0.20
80	8.16	2.04	0.91	0.51	0.33	0.23
90	9.18	2.30	1.02	0.57	0.37	0.26
100	10.20	2.55	1.13	0.64	0.41	0.28
150	15.30	3.83	1.70	0.96	0.61	0.43
200	20.40	5.10	2.27	1.28	0.82	0.57
250	25.50	6.38	2.83	1.59	1.02	0.71
300	30.60	7.65	3.40	1.91	1.22	0.85
400	40.80	10.20	4.53	2.55	1.63	1.13
500	51.00	12.75	5.67	3.19	2.04	1.42
600	61.27	15.30	6.80	3.83	2.45	1.70
700	71.40	17.85	7.93	4.46	2.86	1.98
800	81.60	20.40	9.07	5.10	3.26	2.27
900	91.80	22.95	10.20	5.74	3.67	2.55
1,000	102.00	25.50	11.33	6.38	4.08	2.83

Shaded area is outside the 2-6 feet per second velocity range. If flow is below 2 FPS, air bubbles will be released.

Hydronic cast iron radiator ratings

Column radiators

One Column

Height inches	Sq. Ft per Section	Hydronic 180°F Btuh per Section
20"	1.5	255
23"	1.66	282
26"	2	340
32"	2.5	425
38"	3	510

Two Columns

20"	2	340
23"	2.33	396
26"	2.66	452
32"	3.33	566
38"	4	680
45"	5	850

Three Columns

18"	2.25	383
22"	3	510
26"	3.75	638
32"	4.5	765
38"	5	850
45"	6	1,020

Four Columns

18"	3	510
22"	4	680
26"	5	850
32"	6.5	1,105
38"	8	1,360
45"	10	1,700

Thin Tube radiators

Three Column

Height inches	Sq. Ft. per Section	Hydronic 180°F Btuh per Section
20"	1.75	298
23"	2	340
26"	2.11	359
30"	3	510
36"	3.5	595

Four Column

20"	2.25	383
23"	2.5	425
26"	2.75	468
32"	3.5	595
37"	4.125	701

Five Tube

20"	2.66	452
23"	3	510
26"	3.5	595
32"	4.33	735
37"	5	850

Six Tube

20"	3	510
23"	3.5	595
26"	4	680
32"	5	850
37"	6	1,020

Seven Tube

13"	2.625	447
16 1/2"	3.5	595
20"	4.25	723

Sizing a Pool Heater

Size Pool Heater by Calculating Heat Rise Btu's Required

W = Width   L = Length   SA = Surface area

Pool Dimensions			Gallons in pool		
W	L	SA			
12	24	288	8,617	10,771	12,925
14	28	392	11,729	14,661	17,593
15	30	450	13,464	16,830	20,196
16	32	512	15,319	19,149	22,979
18	36	648	19,388	24,235	29,082
20	40	800	23,936	29,920	35,904
24	45	1,080	32,314	40,392	48,470
25	50	1,250	37,400	46,750	56,100
30	50	1,500	44,880	56,100	67,320
30	60	1,800	53,856	67,320	80,784
30	70	2,100	62,832	78,540	94,248
40	75	3,000	89,760	112,200	134,640
42	75	3,150	94,248	117,810	141,372
42	80	3,360	100,531	125,664	150,797
50	100	5,000	149,600	187,000	224,400
60	100	6,000	179,520	224,400	269,280
60	110	6,600	197,472	246,840	296,208
63	120	7,560	226,195	282,744	339,293

BTU required = Gallons in Pool X 8.34 X Temperature Rise X Hours to raise temperature.

EX 12 x 24-foot pool with an avg. depth of 4 feet at a water temp. of 55°F. We want it to raise it to 78°F in one day or 24 hours.

8,617 gallons x 8.34 (Weight of water) x 23 (Temperature diff) /24 Hours = 68,871 Btuh

Size Pool Heater by Surface Area Btu's Required

Pool Dimensions		Surface Area	Temp Difference	
Width	Length		10°F	20°F
12	24	288	34,560	69,120
14	28	392	47,040	94,080
15	30	450	54,000	108,000
16	32	512	61,440	122,880
18	36	648	77,760	155,520
20	40	800	96,000	192,000
24	45	1,080	129,600	259,200
25	50	1,250	150,000	300,000
30	50	1,500	180,000	360,000
30	60	1,800	216,000	432,000
30	70	2,100	252,000	504,000
40	75	3,000	360,000	720,000
42	75	3,150	378,000	756,000
42	80	3,360	403,200	806,400
50	100	5,000	600,000	1,200,000
60	100	6,000	720,000	1,440,000
60	110	6,600	792,000	1,584,000

Pool Dimensions		Surface Area	Temp Difference	
Width	Length		30°F	40°F
12	24	288	103,680	138,240
14	28	392	141,120	188,160
15	30	450	162,000	216,000
16	32	512	184,320	245,760
18	36	648	233,280	311,040
20	40	800	288,000	384,000
24	45	1,080	388,800	518,400
25	50	1,250	450,000	600,000
30	50	1,500	540,000	720,000
30	60	1,800	648,000	864,000
30	70	2,100	756,000	1,008,000
40	75	3,000	1,080,000	1,440,000
42	75	3,150	1,134,000	1,512,000
42	80	3,360	1,209,600	1,612,800
50	100	5,000	1,800,000	2,400,000
60	100	6,000	2,160,000	2,880,000
60	110	6,600	2,376,000	3,168,000

Source: Energy.Gov

Based on 3 1/2 MPH wind and a rise of 1-1/4°F rise per hour. Temperature difference is the temperature of the water and the ambient air temperature.

For 1 1/2 degree rise per hour multiply above by 1.5

For a 2-deg. rise per hour multiple the above numbers by 2

American Red Cross and FINA (International Swimming Federation) suggests a pool temperature between 77-82 Degrees F

Degree Days and Design Temperatures
Based on 97 1/2% figures
Degree Days / Design Temperatures

ST	Station	Heating Degree Days	Heating Design Temp F
AL	Birmingham	2,551	21
AL	Huntsville	3,070	16
AL	Mobile	1,560	29
AL	Montgomery	2,291	25
AK	Anchorage	10,864	-18
AK	Fairbanks	14,279	-47
AK	Juneau	9,075	1
AK	Nome	14,171	-27
AZ	Flagstaff	7,152	4
AZ	Phoenix	1,765	34
AZ	Tucson	1,800	32
AZ	Yuma	974	39
AR	Fort Smith	3,292	17
AR	Little Rock	3,219	20
AR	Texarkana	2,533	23
CA	Fresno	2,611	30
CA	Long Beach	1,803	43
CA	Los Angeles	2,061	43
CA	Los Angeles	1,349	40
CA	Oakland	2,870	36
CA	Sacramento	2,502	32
CA	San Diego	1,458	44
CA	San Francisco	3,015	38
CA	San Francisco	3,001	40
CO	Alamosa	8,529	-16
CO	Colorado Springs	6,423	2
CO	Denver	6,283	1
CO	Grand Junction	5,641	7
CO	Pueblo	5,462	0
CT	Bridgeport	5,617	9
CT	Hartford	6,235	7
CT	New Haven	5,897	7
DE	Wilmington	4,930	14
DC	Washington	4,224	17
FL	Fort Myers	442	44
FL	Jacksonville	1,239	32
FL	Key West	108	57
FL	Miami	214	47
FL	Orlando	766	38
FL	Pensacola	1,463	29
FL	Tallahassee	1,485	30
FL	Tampa	683	40
FL	West Palm Beach	253	45

State	City	Value 1	Value 2
GA	Athens	2,929	22
GA	Atlanta	2,881	22
GA	Augusta	2,397	23
GA	Columbus	2,383	24
GA	Macon	2,136	25
GA	Rome	3,340	22
GA	Savannah	1,819	27
HI	Hilo	0	62
HI	Honolulu	0	63
ID	Boise	5,809	10
ID	Lewiston	2,542	6
ID	Pocatello	7,033	-1
IL	Chicago (Midway)	6,155	-8
IL	Chicago (O'Hare)	6,639	-4
IL	Chicago	5,882	-2
IL	Moline	6,408	-4
IL	Peoria	6,025	-4
IL	Rockford	6,830	-4
IL	Springfield	5,429	2
IN	Evansville	4,435	9
IN	Fort Wayne	6,205	1
IN	Indianapolis	5,699	2
IN	South Bend	6,439	1
IA	Burlington	6,114	-3
IA	Des Moines	6,588	-5
IA	Dubuque	7,376	-7
IA	Sioux City	6,951	-7
IA	Waterloo	7,320	-10
KS	Dodge City	4,986	5
KS	Goodland	6,141	0
KS	Topeka	5,182	4
KS	Wichita	4,620	7
KY	Covington	5,265	6
KY	Lexington	4,683	8
KY	Louisville	4,660	10
LA	Alexandria	1,921	27
LA	Baton Rouge	1,560	29
LA	Lake Charles	1,459	31
LA	New Orleans	1,385	33
LA	Shreveport	2,184	25
ME	Caribou	9,767	-13
ME	Portland	7,511	-1
MD	Baltimore	4,654	13
MD	Baltimore	4,111	17
MD	Frederick	5,087	12
MA	Boston	5,634	9
MA	Pittsfield	7,578	-3
MA	Worcester	6,969	4
MI	Alpena	8,506	-6
MI	Detroit (city)	6,232	6
MI	Escanaba	8,481	-7
MI	Flint	7,377	1
MI	Grand Rapids	6,894	5
MI	Lansing	6,909	1
MI	Marquette	8,393	-8
MI	Muskegon	6,696	6
MI	Sault Ste. Marie	9,048	-8
MN	Duluth	10,000	-16
MN	Minneapolis	8,382	-12
MN	Rochester	8,295	-12
MS	Jackson	2,239	25
MS	Meridian	2,289	23
MS	Vicksburg	2,041	26
MO	Columbia	5,046	4
MO	Kansas City	4,711	6
MO	St. Joseph	5,484	2
MO	St. Louis	4,900	6
MO	St. Louis	4,484	8
MO	Springfield	4,900	9
MT	Billings	7,049	-10
MT	Great Falls	7,750	-15
MT	Helena	8,129	-16
MT	Missoula	8,125	-6
NE	Grand Island	6,530	-3
NE	Lincoln	5,864	-2
NE	Norfolk	6,979	-4
NE	North Platte	6,684	-4
NE	Omaha	6,612	-3
NE	Scottsbluff	6,673	-3
NV	Elko	7,433	-2
NV	Ely	7,733	-4
NV	Las Vegas	2,709	28
NV	Reno	6,332	10
NV	Winnemucca	6,761	3
NH	Concord	7,383	-3
NJ	Atlantic City	4,812	13
NJ	Newark	4,589	14
NJ	Trenton	4,980	14
NM	Albuquerque	4,348	16
NM	Raton	6,228	1
NM	Roswell	3,793	18
NM	Silver City	3,705	10
NY	Albany	6,875	-1
NY	Albany	6,201	1
NY	Binghamton	7,286	1

State	City	Value	Temp
	Buffalo	7,062	6
	NY(central park)	4,871	15
	NY(Kennedy)	5,219	15
	NY (LaGuardia)	4,811	15
	Rochester	6,748	5
	Schenectady	6,650	1
NY	Syracuse	6,756	2
	Charlotte	3,181	22
	Greensboro	3,805	18
	Raleigh	3,393	20
NC	Winston-Salem	3,595	20
	Bismarck	8,851	-19
	Devils Lake	9,901	-21
	Fargo	9,226	18
ND	Williston	9,243	-21
	Akron-Canton	6,037	6
	Cincinnati	4,410	6
	Cleveland	6,351	5
	Columbus	5,660	5
	Dayton	5,622	4
OH	Mansfield	6,403	5
	Sandusky	5,796	6
	Toledo	6,494	1
	Youngstown	6,417	4
	Oklahoma City	3,725	13
OK	Tulsa	3,860	13
	Eugene	4,726	22
	Medford	5,008	23
	Portland	4,635	23
OR	Portland	4,109	24
	Salem	4,754	23
	Allentown	5,810	9
	Erie	6,451	9
	Harrisburg	5,251	11
	Philadelphia	5,144	14
PA	Pittsburgh	5,053	7
	Reading	4,945	13
	Scranton	6,254	5
	Williamsport	5,934	7
	Providence	5,954	9
RI	Charleston	2,033	27
	Charleston	1,794	28
SC	Columbia	2,484	24
	Huron	8,223	-14
SD	Rapid City	7,345	7
	Sioux Falls	7,839	-11
SD	Bristol	4,143	14
TN	Chattanooga	3,254	18
	Knoxville	3,494	19
	Memphis	3,232	18
	Nashville	3,578	14
	Abilene	2,624	20
	Austin	1,711	28
	Dallas	2,363	22
	El Paso	2,700	24
	Houston	1,396	32
TX	Midland	2,591	21
	San Angelo	2,255	22
	San Antonio	1,546	30
	Waco	2,030	26
	Wichita Falls	2,832	18
	Salt Lake City	6,052	8
UT	Burlington	8,269	-7
VT	Lynchburg	4,166	16
	Norfolk	3,421	22
	Richmond	3,865	17
VA	Roanoke	4,150	16
	Olympia	5,236	22
	Seattle-Tacoma	5,145	26
WA	Seattle	4,424	27
	Spokane	6,655	2
	Charleston	4,476	11
	Elkins	5,675	6
WV	Huntington	4,446	10
	Parkersburg	4,754	11
	Green Bay	8,029	-9
	La Crosse	7,589	-9
WI	Madison	7,863	-7
	Milwaukee	7,635	-4
WI	Casper	7,410	-5
	Cheyenne	7,381	-1
WY	Lander	7,870	-11
	Sheridan	7,680	8

Water Scalding Times		
Temperature °F	1st Degree	2nd/3rd Degree
111.20	5 Hrs.	7 Hrs.
116.60	35 Mins	45 Mins
118.40	10 Mins	14 Mins
122.00	1 Minute	5 Mins
131.00	5 Secs	22 Secs
140.00	2 Secs	5 Secs
149.00	1 Sec	2 Secs
158.00	1 Sec	1 Sec

Outdoor Design Temperatures Canada

Based on 97 ½% figures

Prov	Canadian Location	Design Temperature F
AL	Calgary	-23
AL	Edmonton	-25
AL	Grand Prairie	-37
AL	McMurray	-39
BC	Kamloops	-10
BC	Prince George	-31
BC	Vancouver	19
BC	Victoria	23
MB	Churchill	-39
MB	Dauphin	-26
MB	Winnipeg	-27
NB	Edmunston	-16
NB	Fredericton	-11
NB	Moncton	-7
NB	St. John	-8
NL	Gander	-1
NL	Goose Bay	-25
NL	St. Johns	7
NS	Halifax	5
NS	Yarmouth	9
ON	Hamilton	1
ON	Kenora	-28
ON	London	0
ON	Ottawa	-13
ON	Sault Ste Marie	-15
ON	Timmins	-28
ON	Toronto	-1
PE	Charlottetown	-4
QC	Montreal	-10
QC	Quebec	-12
QC	Sept Iles	-22
QC	Val d'Or	-27
SK	Prince Albert	-35
SK	Regina	-29
SK	Saskatoon	-31
YT	White Horse	-43

Scale forms on the hottest surfaces.

Heating Formulas

Miscellaneous Boiler Information	
1 Watt =	3,415 BTUH
1 Psig =	6.9 bar
Inches Mercury	Psi/.491
Btu/Hour =	Boiler HP x 33,479
	Lb / Hour condensate x 960
1 Boiler HP =	Lbs/Hr * Factor of Evap / 34.5
	34,500 Btuh
	34.5 Lbs Steam/ Hr at 212^{0}F
	34.5 Lb H2O/ Hr
	140 EDR
	about 10-11 ½ square feet of boiler heating surface
	0.069 GPM
	4.14 GPH
	Evaporation of 15.65 kg of water/hr from at 100^{0}C
	8450 kcar/hr
1 HP =	0.746 KW
	746 WATTS
	2,545 BTUH
	1.0KVA
1 Btu =	will raise 1 cubic feet of air 55^{0}F
	will raise 55 cubic feet of air 1^{0}F
	Amount of heat required to raise one pound of water one degree

Heat Transfer Coefficients for still fluids		
Transfer Material	Substance to be heated	Btu/ft^2 hr ^{0}F
Cast Iron	Water	160
Mild Steel	Water	185
Copper	Water	205
Stainless Steel	Water	120
Cast Iron	Air	2.0
Mild Steel	Air	2.5
Copper	Air	3.0

Calculate system pressure required
Divide the elevation of the highest radiator by 2.3 or multiply elevation by 0.43
Add 3-4 psig to that number and that is the pressure needed.

Hydronic Formulas

Fluid Power in HP =	(PSI x GPM) / 1714
Btus =	500* x GPM x Δt
	KW/hr x 3.413
	Kg. cal. x 3.9685
	Watt/hr x 3.413
*500 = 8.33(Weight of a gallon of water) x 60 (minutes)	
Δt = Temperature difference F°	
Calculate pump GPM	Divide heating load in BTUs by 10,000 Btuh for a 20°F delta †
	Divide heating load in BTUs by 15,000 Btuh for a 30°F delta †
Joules =	Btus x 1054.8
	Calories x 4.186
Kg Cal =	Btus x 0.2520
	Joules x 0.0002389
GPM =	$BTUH \div 500 \over \Delta °F$
Δ°F =	$BTUH \div 500 \over GPM$
Typical temperature drop over system =	20-30°F
Typical temperature rise through boiler =	20-30°F
Typical temperature drop over hydronic coil =	10-30°F
BTU/ Output	0.45 Btu/lb. °F @ 68°F
One inch of mercury =	0.433 psi
Hydronic baseboard runs over 30 feet long should be avoided unless using an expansion loop.	
AHRI recommends a minimum velocity of 0.25 Ft/sec for baseboard radiation to avoid laminar flow.	

Heat Output based on Temperature Difference

Flow Rate GPM	Temperature Difference °F			
	5°F	10°F	15°F	20°F
1	2,470	4,940	7,410	9,880
2	4,940	9,880	14,820	19,760
3	7,410	14,820	22,230	29,640
4	9,880	19,760	29,640	39,520
5	12,350	24,700	37,050	49,400
6	14,820	29,640	44,460	59,280
7	17,290	34,580	51,870	69,160
8	19,760	39,520	59,280	79,040
9	22,230	44,460	66,690	88,920
10	24,700	49,400	74,100	98,800

Sizing a Hydronic System Buffer Tanks US Gallons

$$\frac{RT \times (BO - SL)}{\Delta T \times 500}$$

RT – Minimum run time, typically 10-20 minutes
BO – Minimum boiler output
SL – Minimum system load
ΔT – System Temperature drop, typically 20°F

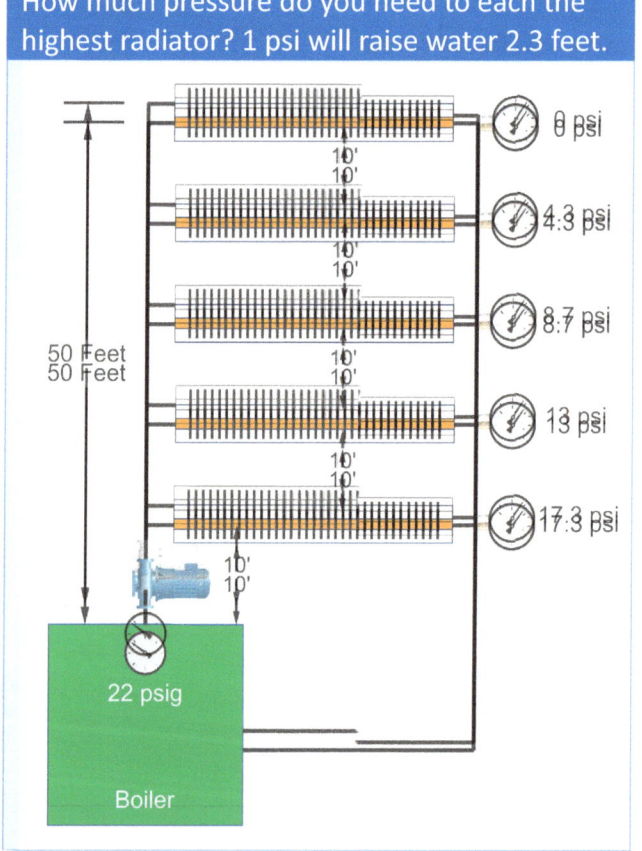

How much pressure do you need to each the highest radiator? 1 psi will raise water 2.3 feet.

Radiant Floor Heat Rules of Thumb

PEX Tubing Size	Maximum Flow GPM	Max. length of tubing run in feet	Typ. # of tubes per manifold
½"	0.575	300	8
5/8"	1	450	12

10-20°F per tube length
Surface temperature 85°F or lower

Max. Pipe Flow Rates based upon 20°F Delta T

Copper Pipe

Pipe Size	Max. GPM	Btuh
½"	1 ½	15,000
¾"	4	40,000
1"	8	80,000
1 ¼"	14	140,000
1 ½"	22	220,000
2"	45	450,000
2 ½"	85	850,000
3"	130	1,300,000

Steel Pipe

Pipe Size	Max. GPM	Btuh
½"	2	12,000
¾"	4	40,000
1"	8	80,000
1 ¼"	16	160,000
1 ½"	22	220,000
2"	50	500,000
2 ½"	80	800,000
3"	140	1,400,000
4"	300	3,000,000
6"	850	8,500,000
8"	1,800	18,000,000
10"	3,500	35,000,000
12"	5,000	50,000,000

PEX Piping

Pipe Size	Max. GPM	Btuh
3/8"	1.2	12,000
1/2"	2	20,000
5/8"	4	40,000
3/4"	6	60,000
1"	9.5	95,000

> A burner should operate for 15 minutes to allow the flame to stabilize before doing air to fuel adjustments.

Expansion Compression Tank Sizing

Size a Compression Tank by Sq. Ft. of Radiation

Recommended Sizing for Compression Tanks

Nominal Capacity Gallons	Sq. ft Radiation
18	350
21	450
24	650
30	900
35	900
35	1100

Compression Tank Sizing Rules of Thumb

One gallon =	23 square feet of radiation
One gallon =	3,500 Btu of radiation.

Compression Tank Sizing Formulas

Closed Tank

$$V_t = V_s \frac{[(V_2/V_1) - 1] - 3\alpha \Delta t}{(P_a/P_1) - (P_a/P_2)}$$

Diaphragm Tank

$$V_t = V_s \frac{[(V_2/V_1) - 1] - 3\alpha \Delta t}{1 - (P_1/P_2)}$$

Definitions

V_t = Volume of compression tank in gallons
V_s = Volume of water in system in gallons
V_1 = Ground water temperature
V_2 = Design heating water temperature (180°F)
$\Delta T = T_2 - T_1$ °F
T_1 = Lower system temperature, typically 40-50°F at fill condition
T_2 = Higher system design temperature, typically 180-220°F.
P_a = Atmospheric pressure (14.7 Psia)
P_1 = System fill pressure Minimum System pressure (Psia)
P_2 = System operating pressure Maximum System pressure (Psia)
α = Linear Coefficient of expansion
  Steel 6.5 x 10⁻⁶
  Copper 9.5 x 10⁻⁶

Use the acceptance factor when choosing the diaphragm tank. The acceptance factor is the amount of space available in the tank.

Sizing Closed Expansion Tank
$V_1 = 40°F = 0.01602$ ft²/lb. (Ground water temperature)
$V_2 = 220°F = 0.01677$ ft²/lb. (Design Temperature)
Vt = tank volume in gallons
α = thermal expansion for steel pipe is 6.5×10^{-6}
Typical system fill pressure 10 Psi
$Vt = Vs \dfrac{[(V2/V1) - 1] - 3\alpha\Delta t}{(P\alpha/P1) - (P\alpha/P2)}$
$Vt = 2000 \dfrac{[(0.01677/0.01602) - 1] - 3(6.5 \times 0.000001) \times 180)}{(14.7/24.7) - (14.7/39.7)}$
$Vt = 2000 \dfrac{0.0468 - 0.00351}{0.595 - 0.370}$
$Vt = 2000 \dfrac{0.0433}{.225}$
Vt = 385 Gallons

Expansion Tank as a Percent of System Volume

Maximum System °F	Type of Expansion Tank		
	Closed	Diaphragm Tank Volume	Diaphragm Acceptance Volume
100	2.21	1.32	0.59
120	3.71	2.21	0.99
140	5.67	3.37	1.51
160	7.87	4.68	2.10
180	10.53	6.27	2.81
200	13.20	7.86	3.52

Sizes based on the following:
Initial Temperature 50°F
Initial Pressure 10 Psig
Maximum operating pressure 30 Psig

Compression Tank Sizing Rules of Thumb*
Steel Piping in building

Entering pressure of 10 psig and maximum pressure of 25 psig
Entering temperature 40°F and maximum water temperature of 225°F

System Capacity in Gallons	Closed Compression tank	Diaphragm Tank
200	39	23
300	58	34
400	77	46
500	96	57
600	116	69
700	135	80
800	154	92
900	173	103
1,000	193	115
1,100	212	126
1,200	231	138
1,300	250	149
1,400	270	160
1,500	289	172
1,600	308	183
1,700	327	195
1,800	347	206
1,900	366	218
2,000	385	229

Compression Tank Sizing Rules of Thumb*
Copper Piping in building

Entering pressure of 10 psig and maximum pressure of 25 psig
Entering temperature 40°F and maximum water temperature of 220°F

System Capacity in Gallons	Closed Compression tank	Diaphragm Tank
200	37	22
300	56	33
400	74	44
500	93	55
600	111	66
700	130	77
800	148	88
900	167	99
1,000	185	110
1,100	204	121
1,200	222	132
1,300	241	143
1,400	260	154
1,500	278	165
1,600	297	177
1,700	315	188
1,800	334	199
1,900	352	210
2,000	371	221

Pumps

Sizing a Circulator GPM
For a 20°F rise, divide boiler output X 10,000
For a 30°F rise, divide boiler output X 15,000
Formula for estimating pump GPM: GPM = BTUH output / (8.33 x 60 x Δt)

Example: Is the existing 40 GPM pump large enough for the new boiler? The new boiler has a rated output of 800,000 with a design temperature rise of 20°F.

800,000 / 10,000 = 80 GPM

The pump is too small for the project. To estimate the capacity boiler the pump could handle, use the following formula:

500* x GPM x Temperature Rise = Btuh

500* x 40 GPM x 20°F Rise = 400,000 Btuh
*500 = 8.33 (weight of 1 gal of water) x 60 (Minutes)

Calculate Pump Head
Measure longest pipe in feet for both supply and return.
Multiply length by 1.5 to calculate fittings and valves.
Multiply by 0.04 - 4' head for each 100' of pipe ensures quiet operation)

For example: 100 feet is longest run (50 ft. supply, 50 ft. return)
100 x 1.5 x .04 = 6 feet of head or 2.6 psig

Estimate Gallons per Minute Water @ various boiler temperature differences		
20°F Delta T	25°F Delta T	30°F Delta T
3.45 GPM per Boiler HP	2.88 GPM per Boiler HP	2.30 GPM per Boiler HP
10,000 Btuh/GPM	12,500 Btuh/GPM	15,000 Btuh/GPM

Radiant heating pumping		
PEX Tubing Size In	Maximum Flow	Maximum Length of Individual Tube Run
½"	0.575 Gpm	300 Feet
5/8"	1 Gpm	450 feet

Required GPM based on Temperature Rise			
Boiler Btuh Output	20°F Rise	25°F Rise	30°F Rise
	Gallons per minute		
50,000	5	4	3
100,000	10	8	7
200,000	20	16	13
300,000	30	24	20
400,000	40	32	27
500,000	50	40	33
600,000	60	48	40
700,000	70	56	47
800,000	80	64	53
900,000	90	72	60
1,000,000	100	80	67
1,100,000	110	88	73
1,200,000	120	96	80
1,300,000	130	104	87
1,400,000	140	112	93
1,500,000	150	120	100
1,600,000	160	128	107
1,700,000	170	136	113
1,800,000	180	144	120

Pump Formulas	
Pump BHP =	$\dfrac{\text{US GPM} \times \text{Head (Ft W.G.)}}{3960 \times \text{Pump Efficiency}}$
$Head_2 =$	$Head_1 \left(\dfrac{RPM\ new}{RPM\ old}\right)^2$
Flow by changing speed	$\dfrac{F2}{F1} = \dfrac{Rpm2}{Rpm1}$
Flow by changing pump impeller	$\dfrac{F2}{F1} = \dfrac{D2}{D1}$
F=Flow in GPM	D=Impeller diameter
Head in feet =	PSI x 2.31/Specific Gravity
Water Source Heat Pump	3.0 GPM per ton @ 10°F ΔT

Typical Pump Types			
Pump Type	GPM	Ft. Head	HP
Circulators	0-150	0-60	¼-5
Close coupled, End Suction	0-2,000	0-400	¼-150
Frame mounted, End Suction	0-2,000	0-500	¼-150
Horizontal split case	0-12,000	0-500	1-500
Vertical Inline	0-2,000	0-400	¼-75

Estimated pump head required based on longest pipe run @ 4 Ft/Head per 100 feet

Longest Pipe Run Ft	Pump Head	Longest Pipe Run Ft	Pump Head
25	1.5	425	25.5
50	3	450	27
75	4.5	500	30
100	6	550	33
125	7.5	575	34.5
150	9	600	36
175	10.5	625	37.5
200	12	650	39
225	13.5	675	40.5
250	15	700	42
275	16.5	725	43.5
300	18	750	45
325	19.5	775	46.5
350	21	800	48
375	22.5	825	49.5
400	24	850	51
425	25.5	875	52.5
450	27	900	54
475	28.5	925	55.5
500	30	950	57
575	32.5	975	58.5
400	24	1000	60

O2 % in Water Based by Temperature

°F	°C	O2 PPM	°F	°C	O2 PPM
50	10	11.1	140	60	4.7
60	15.6	10	150	65.6	4.3
70	21	9.0	160	71.1	3.9
80	27	8.2	170	76.6	3.4
90	32	7.5	180	82.2	2.7
100	37.8	6.9	190	87.8	2
110	43.3	6.3	200	93.3	1.3
120	48.9	5.7	210	98.9	0.6
130	54.4	5.2			

Water Formulas

Water Calculations

To find	Do this
System pressure when boiler is below the radiators.	Highest radiator height x 0.43 to get boiler static pressure.
	Highest radiator height / 2.3 to get boiler static pressure.
Estimate Btus	500 x GPM x Delta T
Delta T	BTU / 500 x GPM
GPM =	BTU / 500 x Delta T
Max water velocity	< 6 FPS @ 200°F
Pump horsepower	HP = (GPM x Total head in feet) / 3960
GPM of water through a pipe.	GPM = 0.0408 x (pipe diameter)² x (water velocity)
Weight of water in a pipe in pounds of water	Lbs of water = 0.34 x pipe length(feet) x (pipe diameter)²
Raise 1 gallon of water 1 degree F =	8.33 Btu
Convert Psig to feet of water	Multiply Psig x 2.307

¹Max flow should be typically 4 FPS. Flow greater than 6 feet per second could cause erosion inside pipe

Water Conversion Factors

Knowing	Multiply by	Desired result
Teaspoon	60	Drops
Tablespoon	3	Teaspoons
Ounce	2	Tablespoons
Cup	16	Tablespoons
Cup	8	Ounces
Pint	2	Cups
PSIG	2.307	Height of water in feet
Quart	2	Pints
US Gallon	8.34	Pounds
US Gallon	0.1338	Cubic Feet
US Gallon	231	Cubic Inches
US Gallon	4	Quarts
US Gallon	8	Pints
US Gallon	16	Cups
US Gallon	3.7853	Liters
US Gallon	0.00379	Cubic Meters
Cu Inch water	0.03613	Pounds
Cu Inch water	0.004329	US Gallons
Cu Inch water	0.576384	Ounces
Cu foot water	62.4	pounds

Lbs. Water	7.49	gallons
	29.92	quarts
	1,728	Cubic Inches
	27.72	Cu. Inches
	0.12	US Gallons
	0.11982	Gallons
	0.45419	Liters
	27.643	Cu. Inches
	0.01603	Cu. Feet
	0.000454	Cu. Meters
One pound of water	27.72	Cu. Inches @ 65°F
	0.12	gallons

Water Density

Temperatures		Density	Density	Specific Volume
°F	°C	Lb/ft3	Lb/gallon	
32	0	63.41	8.48	0.01747
39	3.89	63.42	8.48	0.01602
50	10	63.40	8.48	0.01602
68	20	63.31	8.46	0.01605
86	30	63.12	8.44	0.01609
122	50	62.67	8.38	0.01621
140	60	62.32	8.34	0.01630
158	70	62.01	8.29	0.01638
176	80	61.63	8.24	0.01648
194	90	61.22	8.18	0.01659
212	100	60.78	8.13	0.01671

Water Pressure to Feet Head

Pounds Per Sq. Inch	Feet Head	Pounds Per Sq. Inch	Feet Head
1	2.31	20	46.18
2	4.62	25	57.72
3	6.93	30	69.27
4	9.24	40	92.36
5	11.54	50	115.45
6	13.85	60	138.54
7	16.16	70	161.63
8	18.47	80	184.72
9	20.78	90	207.81
10	23.09	100	230.90
15	34.63		

Feet Head to Water Pressure

Feet Head	Pounds Per Sq. Inch	Feet Head	Pounds Per Sq. Inch
1	.43	60	25.99
2	.87	70	30.32
3	1.30	80	34.65
4	1.73	90	38.98
5	2.17	100	43.31
6	2.60	110	47.64
7	3.03	120	51.97
8	3.46	130	56.30
9	3.90	140	60.63
10	4.33	150	64.96
15	6.50	160	69.29
20	8.66	170	73.63
25	10.83	180	77.96
30	12.99	200	86.62
40	17.32	250	108.27
50	21.65	300	129.93

Percent of heat loss due to scale formation

Scale thickness	Soft carbonate	Hard carbonate	Hard sulfite
1/50"	3.5	5.2	3
1/32"	7	8.3	6
1/25"	8	9.9	9
1/20"	10	11.2	11
1/16"	12.5	12.6	12.6

Want to know how much water is in the system? An inexpensive flow meter like this can attach to a hose you use to fill the system

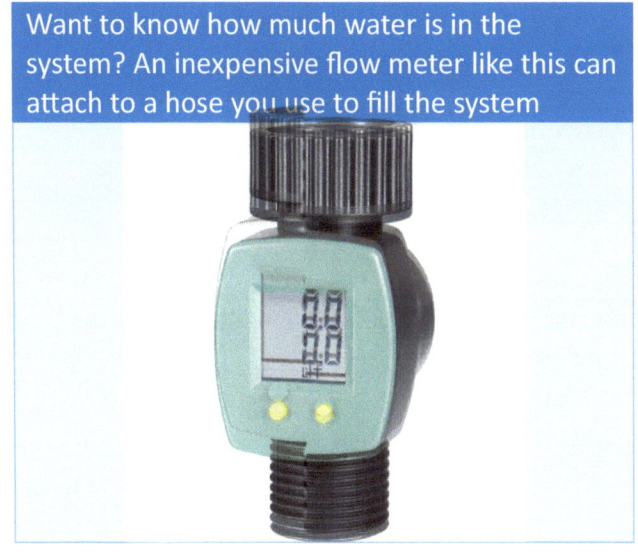

Estimate Hydronic System Volume

The most accurate method is to measure and note the actual pipe sizes in the hydronic loop. This could be done by consulting the building blueprints.

Estimate Hydronic System Volume
Multiply steel compression tank volume by 5-7
35 – 50 gallons per Boiler HP
Pump GPM x 4
Compression tank volume is 20% of system volume
Hot water loop will be about 2/3 the size of a chilled water loop.

A common method for estimating system volume is to use salt because it is easy to test for, very soluble, and inexpensive. The disadvantage to this type of test is that the system must be flushed at the end of the test, wasting water and chemicals. If not, the high chloride levels can be corrosive to the system.

Salt Test Procedure
Fill the system with fresh water.
Circulate and flush the system until the water is clear.
Eliminate all sources of water loss such as bleed, overflow, etc.
Measure the chloride Cl concentration in the system and estimate the system volume.
Add one pound of Table Salt (Sodium Chloride) per 1,000 gallons of estimated volume. This can be added in the pot feeder.
Verify the salt mixes thoroughly.
Allow one hour for the salt to be mixed into the system.
Re-Measure the chloride concentration
Multiply the estimated gallons of water by 76 ppm.
Divide this by the difference (increase) in chloride concentration.
The answer will be the actual system volume

Average System Water Content US Gallons	
Cast Iron Radiation	
Radiator, Large Tube	0.114 gal/ sq. foot
Radiator, Thin Tube	0.056 gal/ sq. foot
Convectors	1.5 Gal/10,000 Btu/Hr @ 200°F
Baseboard	4.7 Gal/10,000 Btu/Hr @ 200°F
Radiation Non-Ferrous	
Convectors	0.64 Gal/10,000 Btu/Hr @ 200°F
Baseboard ¾"	.37 Gal/10,000 Btu/Hr @ 200°F
Fan Coil / Unit Htr.	.2 Gal/10,000 Btu/Hr @ 180°F
100 feet of ¾" copper holds = 2.5 gallons	
100 feet of ¾" copper holds = 4.3 gallons	

Water Pipe Capacity			
Schedule 40 Steel Pipe US			
Gallons per Foot			
Pipe Size Inches	Capacity/ ft	Pipe Size Inches	Capacity/ ft
½"	0.016	3"	0.390
¾"	0.023	4"	0.690
1"	0.040	5"	1.100
1 ¼"	0.063	6"	1.500
1 ½ "	0.102	8"	2.599
2"	0.170	10"	4.096
2 ½"	0.275	12"	5.815
Copper Tubing US			
Gallons per Foot			
	Type K	Type L	Type M
3/8"	0.006	0.007	0.008
½"	0.011	0.012	0.013
5/8"	0.017	0.017	
¾"	0.023	0.025	0.027
1 ½"	0.089	0.092	0.095
2"	0.159	0.161	0.165
2 ½"	0.242	0.248	0.254
3"	0.345	0.354	0.363

> A 1/25" air curtain has the same R value as a four-foot-thick wall of iron.

Glycol

Difference between Freeze and Burst Protection

Burst protection is if your heating system sits dormant (no flow) at temperatures below freezing, putting the pipes in danger of bursting. This sometimes results in a slushy mixture inside the pipes. A slushy mixture is one that contains water, glycol, and frozen ice crystals. Pumping this slushy mixture can damage system components. Consideration for expansion is needed as the mixture expands as it freezes.

Freeze protection is required if your heating system/fluid is going to be pumped at temperatures at or below the freezing point of the fluid. The system must have enough glycol to prevent ice crystals from forming. More glycol is required for freeze protection, keeping the fluid completely liquid, than it does for burst protection, where a slushy mixture is acceptable.

| BTU per Hour with Glycol ||
Glycol Percentage & Type	Formula BTUH =
No Glycol	GPM x 500 x Δt °F
30% E. Glycol @ 68 °F	GPM x 445 x Δt °F
50% E. Glycol @ 32 °F	GPM x 395 x Δt °F
30% P. Glycol @ 68 °F	GPM x 465 x Δt °F
50% P. Glycol @ 32 °F	GPM x 420 x Δt °F

| Propylene Glycol Freeze and Burst Protection |||
Temp F	Freeze Protection % by volume	Burst Protection % by volume
20	18%	12%
10	29%	20%
0	36%	24%
-10	42%	28%
-20	46%	30%
-30	50%	33%

| Propylene Glycol Heat and Flow Correction |||
% of Volume	Heat Transfer	Pump Flow
20%	0.987	1.013
25%	0.978	1.022
30%	0.969	1.032
35%	0.957	1.045
40%	0.944	1.059
45%	0.928	1.077
50%	0.912	1.096
55%	0.893	1.120
60%	0.873	1.145

| Propylene Glycol Pressure Drop Correction |||
% of Volume	140°F Solution	100°F Solution
20%	1.067	1.098
25%	1.078	1.120
30%	1.089	1.141
35%	1.106	1.168
40%	1.122	1.196
45%	1.139	1.228
50%	1.156	1.261
55%	1.172	1.293
60%	1.189	1.326

| Propylene Glycol Freeze and Boiling Points |||
| | Temperature Degrees F ||
Percent of Glycol	Freeze Point	Boiling Point
0	32	212
10	26	212
20	19	213
30	8	216
40	-7	219
50	-28	222
60	<-60	225
70	<-60	230
80	<-60	>230
90	<-60	>230
100	<-60	>230

> A water drip per second equals 8 gallons per day or almost 3,000 gallons per year lost.

Piping Internal Volume of Schedule 40 Pipe

Pipe Size	Square Inches	Pipe Size	Square Inches	Pipe Size	Square Inches
2"	3.36	4"	12.73	8"	51.15
2 1/2"	4.78	5"	19.99	10"	81.55
3"	7.39	6"	28.89	12"	114.80

Standard Nipples & Pipe Sizing Schedule 40

Pipe Size	Outside Diam. in.	Circumference
1/8"	0.405"	1.272"
1/4"	0.540"	1.696"
3/8"	0.675"	2.121"
1/2"	0.840"	2.639"
3/4"	1.050"	3.299"
1"	1.315"	4.131"
1 1/4"	1.660"	5.215"
1 1/2"	1.900"	5.969"
2"	2.375"	7.461"
2 1/2"	2.875"	9.032"
4"	4.500"	14.137"
6"	6.625"	20.813"
8"	8.625"	27.095"
10"	10.750"	33.771"
12"	12.750"	40.054"

Standard Copper Tubing Type K, L, M

Pipe In.	Outside Diameter In.	Circumference
1/2"	0.625"	1.964"
3/4"	0.875"	2.749"
1"	1.125"	3.534"
1 1/4"	1.375"	4.319"
1 1/2"	1.625"	5.105"
2"	2.125"	6.675"
2 1/2"	2.625"	8.246"
3"	3.125"	9.817"
4"	4.125"	12.959"
6"	6.125"	19.242"
8"	8.125"	25.525"
10"	10.125"	31.808"
12"	12.125"	38.092"

> A 1" pipe can transport as many Btus as a 20" round duct.

Equivalent Length
Length of pipe to be added for each fitting

Pipe size in.	Elbow	Gate Valve	Globe Valve
1/2	1.3	0.3	14
3/4	1.8	0.4	18
1	2.2	0.5	23
1 1/4	3.0	0.6	29
1 1/2	3.5	0.8	34
2	4.3	1.0	46
2 1/2	5.0	1.1	54
4	9	1.9	92
6	13	2.8	136
8	17	3.7	180
10	21	4.6	180
12	27	5.5	270
14	30	6.4	310

Number of smaller pipes equal to One larger pipe

Pipe Size	1/2	3/4	1	1 1/4	1 1/2
1/2	1.00	2.27	4.88	10.00	15.80
3/4		1.00	2.05	4.30	6.97
1			1.00	2.25	3.45
1 1/4				1.00	1.50
1 1/2					1.00

Pipe Size	2	2 1/2	4	6
1/2	31.70	52.60	205	620
3/4	14.00	23.30	90	273
1	6.82	11.40	44	133
1 1/4	3.10	5.25	19	68
1 1/2	2.00	3.34	13	39
2	1.00	1.67	6.50	19.60
2 1/2		1.00	3.87	11.70
4			1.00	3.02
6				1.00

Maximum Pipe/Duct Sizes Permitted in Steel Joists

Joist Depth	Round pipe or duct size	Joist Depth	Round pipe or duct size
8"	5"	20"	11"
10"	5"	22"	12"
10"	6"	22"	12"
12"	7"	24"	13"
14"	7"	24"	13"
14"	8"	26"	15"
16"	9"	28"	16"
18"	11"	30"	17"

Pipe Expansion

Thermal Expansion of Piping Material in Inches

Temp Rise °F	Steel	Cast Iron	Copper
32°F - 100°F	0.5	0.5	0.8
32°F - 150°F	0.8	0.8	1.4
32°F - 200°F	1.2	1.2	2.0
32°F - 250°F	1.7	1.5	2.7
32°F - 300°F	2.0	1.9	3.3
32°F - 350°F	2.5	2.3	4.0
32°F - 400°F	2.9	2.7	4.7
32°F - 500°F	3.8	3.5	6.0

Estimated Thermal Expansion of Pipe

Expansion Inches/100 feet from 0°F

Temperature F	Steel	Copper
100°F	0.7	1.1
200°F	1.4	2.2
300°F	2.1	3.3

Expansion cm/10 m from 0°C

Temperature C	Steel	Copper
100°C	0.33	0.61
200°C	0.78	1.23
300°C	1.56	2.46

Calculate Pipe Expansion

If you would like to calculate the expansion or lengthening of a pipe when it has steam or hot water inside, try this formula:

Piping Expansion Calculation

$$E = Constant * Temperature\ rise$$

E = Expansion in inches per 100 feet of pipe
C = Constant

Constant = Coefficient of expansion per 100 Ft pipe

Metal	Constant	Metal	Constant
Steel	0.00804	Cast Iron	0.00780
Wrought Iron	0.00816	Copper or Brass	0.01140

Steel Pipe Expansion

Calculate expansion of 100 feet of Steel Pipe with a temperature rise 50°F to 180°F

$$E = 0.00804 * (180 - 50)\ Temp\ rise$$
$$1.045" = 0.00804 * 130$$

The expansion is 1.045" for every 100 feet of steel pipe.

Copper Expansion

Calculate expansion of 100 feet of copper tubing with a temperature rise 50°F to 180°F

$$E = 0.01140 * (180 - 50)\ Temp\ rise$$
$$1.482" = 0.01140 * 130$$

The expansion is 1.482" for every 100 feet of copper tubing.

Don't forget about expansion

Copper expands 50% more than black iron. Be sure there is room for the copper to expand.

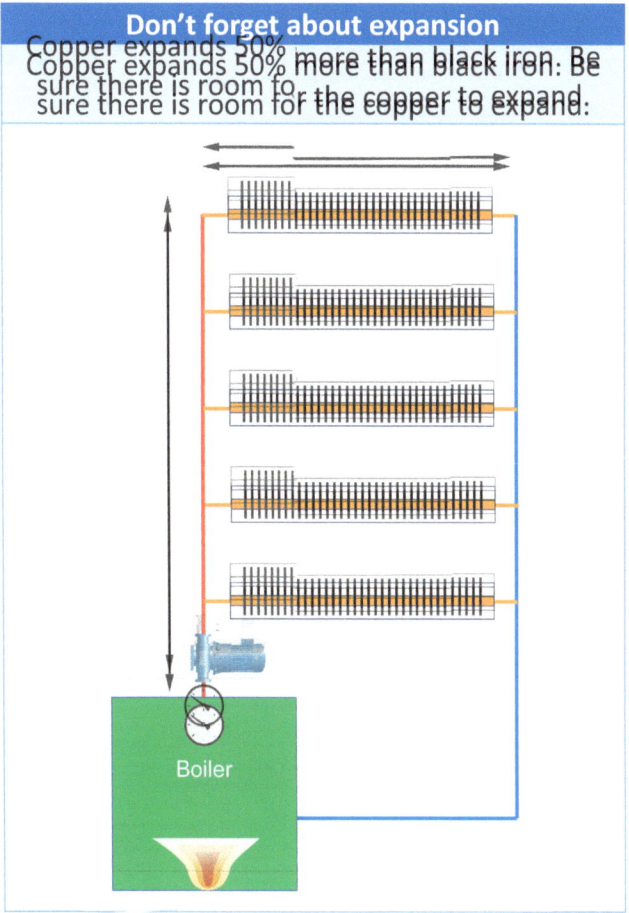

Linkage-less burners are 5-6% more efficient than burners with linkages, according to Honeywell.

Velocity Calculation

Typical Hydronic Velocity = 2 -4.5 Feet per second in occupied areas. Slightly higher in unoccupied areas. Flows above 6 feet per second could erode copper. Flows below 2 could allow air to be trapped in the piping and cause air locked systems.

To find Fluid Velocity	
Feet per Second =	$\dfrac{0.408 \times GPM}{\text{Pipe Diameter Inches}^2}$
Gallons per minute GPM =	$\dfrac{.3208 \times GPM}{\text{Pipe Diameter Inches}^2}$
	Pipe Diameter Inches$^2 \div$ FPS

To find Fluid Velocity

$$\text{Feet per Second} \frac{0.408 \times GPM}{(\text{Pipe Diameter Inches})^2}$$

$$GPM = (\text{Pipe Diameter Inches})^2 \div FPM$$

This table will allow you to calculate the capacity in gallons per minute at different velocities.

For example, if you have a 2" pipe and want to keep the velocity at 4 feet per second, the pipe will be capable of running 54 gallons per minute.

Feet per second FPS to Miles per Hour MPH			
FPS	MPH	FPS	MPH
1	0.68	6	4.09
2	1.37	7	4.77
3	2.05	8	5.45
4	2.73	9	6.13
5	3.40	10	6.82

Velocity		
Knowing	Multiply by	To Get
Ft/Sec	60	Ft/Min
Ft/min	0.01139	Miles/Hr
	0.01667	Ft/Second
Cu Ft/ Min	0.1247	Gal/Sec
Cu Ft/Sec	448.8	Gal/min
Miles/Hr	88	Ft /Min
Gal/minute	0.002228	Cu Ft/Sec

Pipe GPM @ Various Velocities Feet per Second					
Pipe Size Inches	FPS = Feet per Second				
	2 FPS	3 FPS	4 FPS	5 FPS	6 FPS
	Gallons per Minute				
1/2"	2	4	5	12	22
3/4"	4	6	8	21	38
1"	7	10	14	35	62
1 1/4"	12	18	24	60	108
1 1/2"	16	18	33	81	147
2"	27	24	54	135	242
3"	59	89	118	296	533
4"	102	153	204	510	918
6"	231	347	463	1,157	2,082
8"	400	600	800	2,000	3,599
10"	631	946	1,261	3,153	5,675
12"	895	1,343	1,791	4,476	8,058

Convert Flow (GPM) to Velocity (FPS)						
	Pipe Diameter in Inches					
	2	4	6	8	10	12
GPM	FEET PER SECOND					
5	0.51	0.13	0.06	0.03	0.02	0.01
10	1.02	0.26	0.11	0.06	0.04	0.03
15	1.53	0.38	0.17	0.10	0.06	0.04
20	2.04	0.51	0.23	0.13	0.08	0.06
30	3.06	0.77	0.34	0.19	0.12	0.09
40	4.08	1.02	0.45	0.26	0.16	0.11
50	5.10	1.28	0.57	0.32	0.20	0.14
60	6.12	1.53	0.68	0.38	0.24	0.17
70	7.14	1.79	0.79	0.45	0.29	0.20
80	8.16	2.04	0.91	0.51	0.33	0.23
90	9.18	2.30	1.02	0.57	0.37	0.26
100	10.20	2.55	1.13	0.64	0.41	0.28
150	15.30	3.83	1.70	0.96	0.61	0.43
200	20.40	5.10	2.27	1.28	0.82	0.57
250	25.50	6.38	2.83	1.59	1.02	0.71
300	30.60	7.65	3.40	1.91	1.22	0.85
400	40.80	10.20	4.53	2.55	1.63	1.13
500	51.00	12.75	5.67	3.19	2.04	1.42

Carbon monoxide CO has a five-hour life in a body.

Pipe Insulation

Avg. Loss from Insulated pipe		
Pipe Size Inches	Insulation Thickness	175°F
½	1"	0.150
¾	1"	0.172
1	1"	0.195
	1 ½"	0.165
1 ¼"	1"	0.250
	1 ½"	0.170
1 ½"	1"	0.247
	1 ½"	0.205
2"	1"	0.290
	1 ½"	0.235
	2"	0.200
2 ½"	1"	0.330
	1 ½"	0.265
	2"	0.225
3"	1"	0.385
	1 ½"	0.305
	2"	0.257
4"	1"	0.470
	1 ½"	0.370
	2"	0.308
Btu/ Linear foot @ 70 Deg F		

Heat Losses from Uninsulated Horizontal Pipe

Steel Pipe		Copper	
Pipe Size In.	180°F Water	Pipe Size In.	180°F Water
½	60	½	33
¾	73	¾	45
1	90	1	55
1 ¼	112	1 ¼	66
1 ½	126	1 ½	77
2	155	2	97
2 ½	185	2 ½	117
3	221		
4	279		
Btu/ Linear foot @ 70 Deg F			

Heat loss in Btuh per foot of Uninsulated. Schedule 40 Steel Pipe

Subtract ambient air from the temperature inside pipe.

Deg. F. Difference	Pipe Size in Inches			
	¾	1	1 ¼	1 ½
100	68	82	107	113
120	85	104	127	142
140	104	127	155	173
160	125	152	186	213
180	146	176	217	243
200	171	206	251	282
225	199	243	297	334
250	233	284	347	389
275	266	326	398	447
300	304	372	455	510
Deg. F. Difference	Pipe Size in Inches			
	2	2 1/2	3	4
100	138	163	243	337
120	175	206	308	427
140	212	251	375	521
160	256	301	451	626
180	297	351	522	725
200	346	408	622	850
225	410	483	726	1009
250	478	563	849	1180
275	550	649	978	1360
300	628	742	1140	1557

Water transfers heat 24 times faster than air.

Combustion

Calculate CO Air Free from CO Readings

"As Measured" CO Reading

O2 %	25	50	75	100
0	25	50	75	100
1.0	26	53	79	105
2.0	28	55	83	111
3.0	29	58	88	117
4.0	31	62	93	124
5.0	33	66	98	131
6.0	35	70	105	140
7.0	38	75	113	150
8.0	41	81	122	162
9.0	44	88	132	176
10.0	48	96	144	192
11.0	52	106	158	211
12	59	117	176	235
13	66	132	198	263
14	76	151	227	303
15	88	177	266	354

"As Measured" CO Reading

O2 %	125	150	175	200
0	125	150	175	200
1.0	131	158	184	210
2.0	138	166	194	221
3.0	146	175	204	234
4.0	155	186	218	247
5.0	164	197	230	263
6.0	175	210	245	281
7.0	188	225	263	301
8.0	203	243	284	324
9.0	220	263	307	351
10.0	240	288	336	383
11.0	264	317	369	422
12	294	353	411	470
13	331	397	463	529
14	379	454	530	606
15	443	531	620	708

CO Air Free limit in flue is 400 PPM

Perfect Combustion

Amount of air needed for perfect efficiency

Fuel	Combustion Air
Propane	23.5 ft²
Natural Gas	10 ft²

Percent of Air that will permit combustion

	Min % of Air	Max % of Air
Natural Gas	04	247
Oil	30	173
Coal Pulverized	8	425

Fuels Ignition Temperatures

Fuel	Ignition Temp °F	Fuel	Ignition Temp °F
Butane-N	760	Gasoline	735
CO	1,170	Hydrogen	1,095
Coal	850	Kerosene	500
#2 Fuel Oil	600	Natural Gas	1,000
#6 Fuel Oil	765	Propane	875

Produced When One Cu. Ft. of Gas is Burned

One Cubic Foot of Gas Burned Produces	8 Cubic feet of nitrogen
	2 Cubic feet of water vapor
	1 Cubic Foot of Nitrogen

Flue Gas Dewpoint Temperatures

% O2	Dewpoint °F	% O2	Dewpoint °F
3%	133	7%	123
4%	131	8%	122
5%	130	9%	118
6%	128	10%	116

Carbon Monoxide (CO) Exposure Effects

CO PPM	Effects
200	Slight Headache, Tiredness, Dizziness, nausea after 2-3 hours.
400	Frontal Headaches 1-2 Hrs.; Life Threatening After 3 Hours.
800	Dizziness, Nausea & Convulsion within 45 minutes. Unconsciousness within 2 hours. Death Within 2-3 Hours.
1,600	Headache, Dizziness & Nausea within 20 minutes. Death within 1 Hour
3,200	Headache, Dizziness & Nausea within 5-10 minutes. Death within 30 minutes
6,400	Headache, Dizziness & Nausea within 1-2 minutes, Death within 10-15 minutes
12,800	Death Within 1-3 Minutes

Carbon monoxide dangers

Negative condition in boiler room If you look at the pictures below, the top picture is a normal boiler room, and the combustion air opening allows the boiler to operate safely. The lower picture shows how a boiler room can go negative allowing the dangerous flue gases to be pulled from the chimney and enter the boiler room and the building. It only takes -0.02" WC negative pressure to pull the flames from a burner. Exhaust fans in a boiler room could be dangerous.

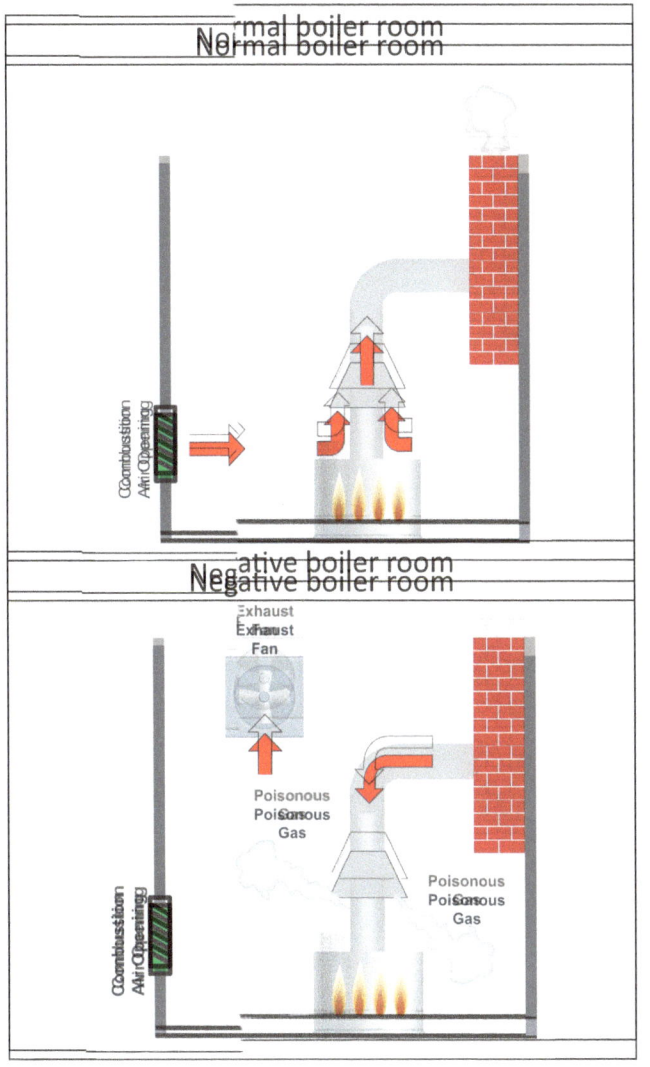

Typical Combustion Readings for Category I boilers

	Natural Gas		#2 oil
	Atmos	Power	Power
O2 %	7-9%	3-6%	4-7%
Stack temp	325 to 500°F	275 to 500°F	325 to 600°F
In. WC Draft	-.0" to -.04"	-.02" to -.04"	-.04" to -.06"
CO PPM	≤100		

Effects of Boiler Draft When it is Too Low or High

Draft too low	Draft too high
Excessive heat in boiler	Elevated flue temperatures
Flue gas spillage into room	Affects boiler efficiency
Could allow Carbon Monoxide to form	Could allow Carbon Monoxide to form
Could cause flame impingement	Could cause flame impingement
Possible damage to burner	

Excess air percentages

Parts air/gas	Excess air %	Parts air/gas	Excess air %
10	0%	16	60%
11	10%	17	70%
12	20%	18	80%
13	30%	19	90%
14	40%	20	100%
15	50%		

Oxygen corrosion is twice as corrosive @ 86°F than it is @122°F in water.

Flame Safeguard Flame Signals

Cad Cell resistance	Less than 1,000 ohms when running
Thermocouple	25 mV minimum
Powerpile thermocouple	750 mV
Flame sensor	3 microamps when meter is connected in series with flame sensor and pilot flame is present
Hot surface ignitor	40 to 80 Ohms typical
	4 to 4.5 amps when lit

Fireye Flame Signals

Model	Average Flame Signal
UVM	4.0-5.5 VDC
TFM	14-17 VDC
D-10/20/30	16-25 VDC
E-100/ 110	20-80 VDC
E-100/E110 with EPD Programmer	4-10 VDC
M Series II	4-6 VDC
Micro M Series	4-10 VDC
Micro M Series w Display	20-80 VDC

Honeywell Flame Signals

Model	Average Flame Signal
RA890	2-6 µA DC
R4795	2-6 µA DC
R7795	2-6 µA DC
R4140	2-6 µA DC
R4150	2-6 µA DC
BC7000	2-6 µA DC
RM7890	1.25-5 VDC
RM7895	1.25 VDC
RM7840	1.25 VDC
RM7800	1.25 VDC
S8600	1-5 µA DC

Typical Flame Signals

Flame Detector	Average Flame Signal
Flame Rod	2-5 µA DC
Flame Rod Self-Checking	1 ¼ - 2 ½ µA DC
Photocell	2-5 µA DC
Photocell Self-Checking	1 ¼ - 2 ½ µA DC
Ultraviolet	2-6 µA DC
Infrared	2 1/4-5 µA DC
Thermocouple Q309	30mV
Thermopile Q313	750 mV
Please verify with manufacturer	

Clocking a Gas Meter

Clocking a gas meter verifies the actual firing rate of the boiler. To properly "clock" a meter, be sure your boiler is the only apparatus firing. The boiler should also be at high fire. A way to check this is to shut off all the items in the boiler room and observe the gas meter. If the hands on the dial are still moving, there are other items consuming gas. It is sometimes difficult to perform this task in a commercial building, as there could be many items operating simultaneously.

Some commercial gas meters require you to compensate for the temperature as well as the pressure of the gas. When the meter was calibrated, it was at a certain temperature and gas pressure at the factory. The meter above was calibrated using 60^0F gas. The field conditions may be different.

How to Clock the Meter
Once you are sure that there are no other appliances using gas, start your unit and assure that it is firing at full rate. Count the number of revolutions the most sensitive dial on the gas meter makes in one minute. Most natural gas has a heating value of 1,000 BTU/cubic foot. Let us assume that our most sensitive dial is ½ cubic feet per revolution.

A. Count the revolutions the ½ cubic foot dial makes in one minute.
B. Multiply the revolutions by 30,000 to obtain the firing rate in Btu's/ Hr

For example, the ½ cubic foot dial made 3.2 revolutions in one minute. The boiler is firing at 3.2 revolutions x 30,000 BTU/revolution = 96,000 BTUH. If you find that the heating value is different from 1,000 Btu's per cubic foot, you will have to make an adjustment. For example, if the heating value is 1,050 BTU/ cubic foot, you would need to adjust your final reading. 1,050 BTU/ Cubic foot (Actual BTU) divided by 1,000 Btu/ Cubic foot (This was assumed to be the BTU content) = 1.05. Therefore, to recalculate the new rate, we would multiply 96,000 Btuh (From above) x 1.050 = 100,800 Btu/ HR. This is the actual firing rate of the appliance.

The 30,000 calculation only works with ½ cubic foot dial. For other size dials, see below.

Remember our basic formula is number of revolutions x factor below = BTU/ Hr. This is based on 1,000 BTU/ Cubic Foot.

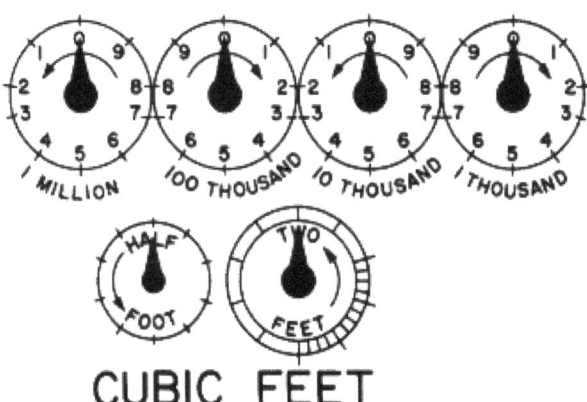

NOTE: To get a more accurate reading, it is better to allow the test to be done for a longer time. I would recommend 5 minutes. You would then divide the reading by 5 to get the average. The chart below features different timing for the dials, up to five minutes. For example, if the 5 "Cubic Feet per Revolution" dial made two revolutions in five minutes, your firing rate would be as follows:

60,000 x 2 = 120,000 Btuh

Multiplying Factor for Gas Meter		
	1 Minute	2 Minute
Cu. Ft / Revolution	BTUH	
½	30,000	15,000
1	60,000	30,000
2	120,000	60,000
5	300,000	150,000
	3 Minute	5 Minute
Cu. Ft / Revolution	BTUH	
½	7,500	6,000
1	15,000	12,000
2	30,000	24,000
5	75,000	60,000
Based on 1,000 Btu per cubic foot of gas		

Clocking a Gas Meter Option 2

Start the boiler; making certain that no other gas-fired appliance is operating. Measure the amount of time it takes for the smallest dial to make one complete revolution. In the above dials, the ½ cubic foot dial is the timing dial.

Refer to a natural gas timing chart under ½ cubic foot column and see what the input is to your boiler.

Check and compare the calculated input with the input rating on the heating unit data plate. If the unit is under-fired or over-fired by more than 10%, check the gas pressure to the unit and adjust as necessary. (For example, the unit being tested takes 29 seconds for the ½ cubic foot dial to make one complete revolution. Using the chart, this translates to 62 cubic feet per hour. Based upon the assumption that one cubic foot of natural gas has 1,000 BTU's (

You will get a better reading by allowing the dial to rotate several times and dividing the total by the number of revolutions to get an average. In the above example, if it took 1 minute, 27 seconds or 87 seconds to make three revolutions, our average input would be 29 seconds.

Natural Gas Timing Chart in Cubic Feet/Hour

Seconds for one revolution	1/2 Cu Ft	1 Cu Ft	2 Cu Ft	5 Cu Ft
10	180	360	720	1,800
11	164	327	655	1,636
12	150	300	600	1,500
13	138	277	554	1,385
14	129	257	514	1,286
15	120	240	480	1,200
16	112	225	450	1,125
17	106	212	424	1,059
18	100	200	400	1,000
19	95	189	379	947
20	90	180	360	900
21	86	171	343	857
22	82	164	327	818
23	78	157	313	783
24	75	150	300	750
25	72	144	288	720
26	69	138	277	692
27	67	133	267	667
28	64	129	257	643
29	62	124	248	621
30	60	120	240	600
31	58	116	232	581
32	56	113	225	563
33	55	109	218	545
34	53	106	212	529
35	51	103	205	514
36	50	100	200	500
37	49	97	195	486
38	47	95	189	474
39	46	92	185	462
40	45	90	180	450
41	44	88	176	440
42	43	86	172	429
43	42	84	167	420
44	41	82	164	410
45	40	80	160	400
46	39	78	157	391
47	38	77	153	383
48	37	75	150	375
49	37	73	147	367
50	36	72	144	360
51	35	71	141	353
52	35	69	138	346
53	34	68	136	340
54	33	67	133	333
55	33	65	131	327
56	32	64	129	321
57	32	63	126	316
58	31	62	124	310
59	31	61	122	305
60	30	60	120	300
62	29	58	116	290
64	28	56	112	281
66	27	55	109	273
68	26	53	106	265
70	26	51	103	257
72	25	50	100	250
74	24	48	97	243
76	24	47	95	237
78	23	46	92	231
80	22	45	90	225

Gas Temperature Correction Factor

Gas °F	Meter Calibration Temperature				
	60°F	65°F	68°F	70°F	72°F
0	1.130	1.141	1.148	1.152	1.157
5	1.118	1.129	1.135	1.140	1.144
10	1.106	1.117	1.123	1.128	1.132
15	1.095	1.105	1.112	1.116	1.120
20	1.083	1.094	1.100	1.104	1.108
25	1.072	1.083	1.089	1.093	1.097
30	1.061	1.071	1.078	1.082	1.086
35	1.051	1.061	1.067	1.071	1.075
40	1.040	1.050	1.056	1.060	1.064
45	1.030	1.040	1.046	1.050	1.053
50	1.020	1.029	1.035	1.039	1.043
55	1.010	1.019	1.025	1.029	1.033
60	1.000	1.010	1.015	1.019	1.023
65	0.990	1.000	1.006	1.010	1.013
70	0.981	0.991	0.996	1.000	1.004
75	0.972	0.981	0.987	0.991	0.994
80	0.963	0.972	0.978	0.981	0.985
85	0.954	0.963	0.969	0.972	0.976
90	0.945	0.955	0.960	0.964	0.967
95	0.937	0.946	0.951	0.955	0.959
100	0.929	0.938	0.943	0.946	0.950

NOTES:
On a commercial gas meter, you may have to calculate a pressure and/or temperature

correction factor. You will need to contact the local gas company for this factor.

Gas Pressure Correction Factor

Actual Meter Pressure (psi)	Meter Base Pressure			
	4 oz or 7" w.c.	8 oz or 14" w.c.	10 oz or 17.5" w.c.	1 Psi or 28" w.c.	2 psi or 56" w.c.
0	0.983	0.966	0.958	0.935	0.878
¼	1.0	0.983	0.975	0.951	0.893
½	1.017	1	0.992	0.968	0.909
5/8	1.026	1.008	1	0.976	0.916
1	1.051	1.034	1.025	1.000	0.939
2	1.119	1.101	1.093	1.066	1.000
3	1.188	1.168	1.158	1.130	1.061
4	1.256	1.235	1.225	1.195	1.122
5	1.324	1.302	1.291	1.260	1.183
6	1.392	1.369	1.358	1.325	1.244
7	1.461	1.436	1.424	1.390	1.305
8	1.529	1.503	1.491	1.455	1.366
9	1.597	1.570	1.557	1.520	1.427
10	1.666	1.638	1.624	1.584	1.488

Water Treatment

Hydronic system water treatment is trickier than steam systems because of the different materials in the system. If you have glycol in the water, now you added a different set of parameters. I suggest installing a water meter on the makeup water pipe. The meter will let the owner know if water is leaking from the system.

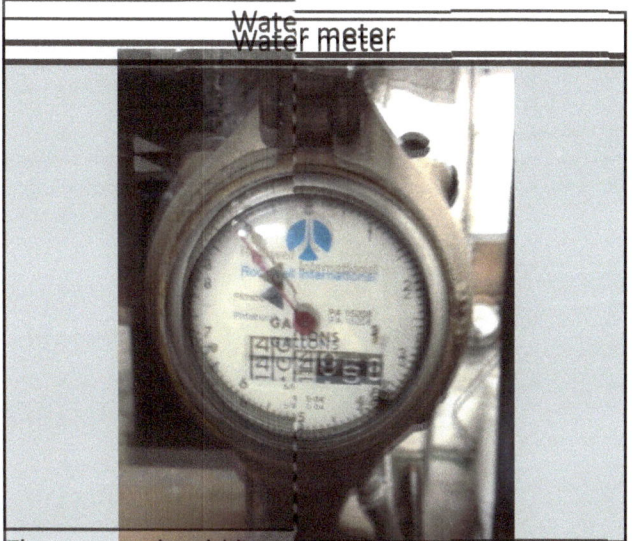

Water meter

The owner should hire a water treatment professional for their recommendations. Be sure to have a list of all the metals in the system. I had an aluminum boiler fail in a few years because the water treatment technician thought it was copper. *Don't buy some online water treatment*.

Many of the newer hydronic components have tighter tolerances than previous ones and that could lead to plugged pipes, circulators, or valves, or strainers. Increased filtration should be installed.

This chemical feeder is used to introduce water treatment chemicals into the system. Some are available with a strainer inside to catch some of the junk inside the water.

> The radiator heat output is 40% radiation and 60% convection.

Pump plugged with rust and dirt

Boiler pH is typically 7-9

pH Readings		
PH Readings	Compare to	Acid/ Alkaline
0	Battery Acid	ACID
1	Hydrochloric Acid	ACID
2	Lemon Juice, Vinegar	ACID
3	Grapefruit /Orange Juice	ACID
4	Acid Rain /Tomato Juice	ACID
5	Black Coffee	ACID
6	Urine / Saliva	ACID
7	Neutral	
8	Sea Water	ALKALINE
9	Baking Soda	ALKALINE
10	Milk of Magnesia	ALKALINE
11	Ammonia	ALKALINE
12	Soapy Water	ALKALINE
13	Bleach / Oven Cleaner	ALKALINE
14	Drain Cleaner	ALKALINE

Why do boiler maintenance?

There are two reasons for performing maintenance on a boiler: safety and longevity. The odds of a boiler accident drop when maintenance is done. The second reason for doing maintenance is longevity; boilers which are maintained last much longer than ones which are not. The leading cause of boiler failures is improper maintenance. Check with the manufacturer of your boiler for their requirements. The following is a generic one I use for most hydronic boilers.

Daily Tasks
Look around the boiler room for unsafe conditions. Maintain 30" clearance around the boilers or water heaters.
Look for signs of flame rollout from boiler or water heater.
Look for water leaks around the boiler and piping. The leaks should be repaired ASAP.
Look for rust under the boiler or water heater draft diverter. This could signify back drafting of the flue gases.
Sniff for natural gas leak. Repair gas leaks asap.
Watch and listen to the blower or induced draft motor. Failing motor bearings may be louder when motor starts or stops.
Watch the firing rate / modulating motor. Verify it works smoothly without binding.
Check chemical feed and softener system operation.
Visually inspect burner flame & verify the flame is not impinging on the metal surfaces.
Be sure the combustion chamber is not covered in soot.
Check the boiler flue draft. It should be slightly negative, in most instances on Category I boilers.
Verify relief valve isn't leaking.
Check boiler pressure gauge. The pressure should be around 12 psi for a 2-story building.
Check stack thermometer when boiler is running. If its higher, this could indicate a dirty fire side or scale on water side.
Verify pipe insulation is intact & no evidence of a leak. Wet insulation should be changed.
Inspect stack. Look inside when boiler is NOT firing to see if soot is forming in the flue.

Verify flue pipe is intact, no deterioration or blockage, or rust holes. Verify it's connected properly and pitched up to the chimney.
Be sure the combustion air openings are free and clear.

Weekly Tasks
Check water meter on the makeup water pipe. Note weekly usage and investigate if water usage increases.
Check supply levels for water treatment and water softener. Be sure you have enough supplies to last the week.
Check boiler/burner light bulbs. Replace burnt out bulbs.
Regenerate water softener if used.
Check fuel train for leaks. Look at window in gas valve actuator for hydraulic fluid leak. Replace the actuator if leaking.

Monthly Tasks
Check gauge glass for damage, wear, or etching. Replace if you notice those things.
Check the linkages on gas train components. Look at linkages and ball joints for wear or slippage.
Check gas train venting. Verify gas train components are properly vented and the vents are connected

Twice Yearly Tasks
Power burner. If the boiler has a power burner, it should be checked twice per year and the fuel to air ratio should be tested and adjusted.

Annual Tasks
Look at fireside of boiler for water leaks. Check refractory for cracks and missing pieces. Repair or replace missing or damaged refractory. NOTE: If refractory gets wet, it should be thoroughly dried before exposing it to flame. Entrained moisture can cause it to break apart when heat is applied.
Open low water cutoffs and inspect probes and floats. Replace any defective or worn components. Remember the system is filled with water and you could get wet or burned.
Open wye strainers. Clean any sludge buildup and clean screens
Check burner blower motor. Verify it is clean and operates without excess vibration.

Check and adjust air to fuel ratio. Use a calibrated combustion analyzer to adjust the air to fuel ratio. This should be a task done by someone trained on boilers.
Leak test the gas train.
Check flame safeguard according to the manufacturer.
Test the relief valve.
Check all wiring terminations. Verify all the screws are snug on the wiring terminals. Be sure the power is off when doing this.
Check boiler base on atmospheric boiler. Look for missing or cracked refractory or damage to the base. Do not operate until it is repaired.
Check burners. Clean if dirty.

To get 1,000,000 Btu's you need the following:
1,000 Cu ft of natural gas
10 Therms of natural gas
1 Mcf or dekatherm of natural gas
10.92 gallons of propane
8 gallons of gasoline
7.19 gallons of #2 fuel oil
293.083 Kwh of electricity
29.3 boiler horsepower

One part per million or PPM=

- 1 day in 2,739 years
- 1 Interception in 7,962 football games.
- 1 inch in 16 miles.
- 1 drop of vermouth in 80 fifths of gin.

Combustion Air

Combustion Air Openings

One Opening

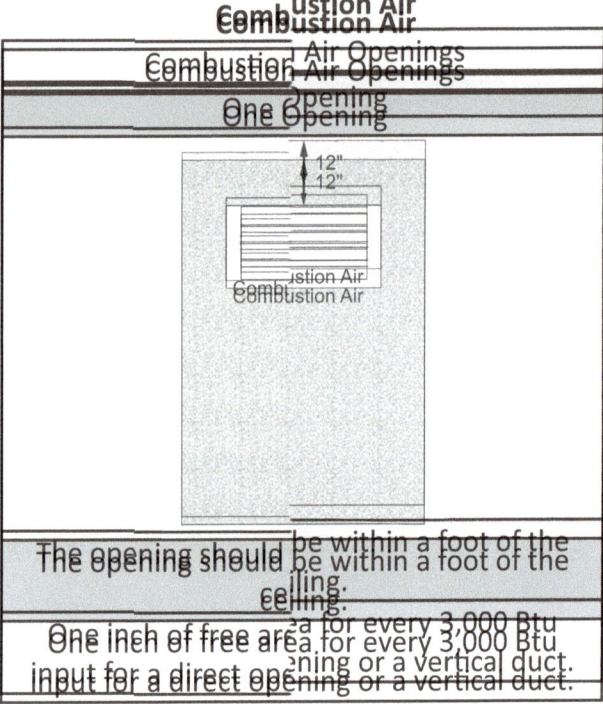

The opening should be within a foot of the ceiling.

One inch of free area for every 3,000 Btu input for a direct opening or a vertical duct.

Two Openings

One opening within a foot of the floor and one within a foot of the ceiling.

One inch of free area for every 4,000 Btu input of the boiler for a direct opening or a vertical duct.

One inch of free area for every 4,000 Btu input of the boiler for a direct opening or a vertical duct.

One inch of free area for every 2,000 Btu input of the boiler for a horizontal duct.

Btuh Capacity of Direct Vent Metal Combustion Air Louvers

Height Inches	Btu Capacity Width in inches		
	12	24	36
12	432,000	864,000	1,296,000
18	648,000	1,296,000	1,944,000
24	864,000	1,728,000	2,592,000
30	1,080,000	2,160,000	3,240,000
36	1,292,000	2,592,000	3,888,000
42	1,512,000	3,024,000	4,536,000
48	1,728,000	3,456,000	5,184,000
60	2,160,000	4,320,000	6,480,000

Based on direct connect metal louvers with 75% free area and 4,000 Btu per inch of free space.

Typical Free Area Estimate for Various Openings and Louvers

Opening Type	Estimated A$_K$ or Free Area
Metal Louver	75% Free Area
Wooden Louver	50% Free Area
Metal Mesh Screen	98% Free Area
Motorized Combustion Air Dampers	95% Free Area

A$_K$ or Area Factor is the actual free area of the grill.

Volume of Air

Degrees F	Cubic ft in 1 lb.	Degrees F	Cubic ft in 1 lb.
0	11.58	130	14.85
32	12.38	140	15.1
40	12.58	150	15.35
50	12.84	160	15.60
60	13.14	170	15.85
70	13.34	180	16.1
80	13.59	200	16.6
90	13.82	210	16.86
100	14.09	212	16.91
110	13.34	220	17.11
120	14.59		

Water flows at speed up to 4 miles per hour in a hydronic system.

Mechanical Combustion Air

0.35 CFM per 1,000 Btuh
International Mechanical Code
Fan operation should be verified before starting the boiler

Btuh	CFM	Btuh	CFM
50,000	18	750,000	263
75,000	26	800,000	280
100,000	35	850,000	298
150,000	53	900,000	315
200,000	70	950,000	333
250,000	88	1,000,000	350
300,000	105	1,500,000	525
350,000	123	2,000,000	700
400,000	140	2,500,000	875
450,000	158	3,000,000	1,050
500,000	175	3,500,000	1,225
550,000	193	4,000,000	1,400
600,000	210	4,500,000	1,575
650,000	228	5,000,000	1,750
700,000	245		

Things to check on combustion air openings

- Are the opening sized correctly?
- Are opening free and clean and not blocked or covered?
- Louver screen opening sizes should be no less than ¼" and no larger than ½"
- Verify water pipes are not in front of combustion air openings.
- Verify interlock works on the combustion air damper to assure damper is open before starting burner.
- Verify the boiler room is not allowed to go negative from an exhaust fan or a leaking return duct on air handling units. It only takes -0.03" WC negative pressure to pull the flue gases from the flue or atmospheric burner.

Indoor Air

50 cubic feet of volume for each 1,000 Btuh
Add all the gas fired equipment in the room:
Divide total by 50: This is how much volume is needed for the appliances.
Measure ceiling height.
The columns to the right show how much area is required based on the ceiling height

Total MBtuh	Cu Ft volume needed	Ceiling Height in Feet			
		10	15	20	25
		Area needed			
50	1,000	100	67	50	40
100	2,000	200	133	100	80
200	4,000	400	267	200	160
300	6,000	600	400	300	240
400	8,000	800	533	400	320
500	10,000	1,000	667	500	400
600	12,000	1,200	800	600	480
700	14,000	1,400	933	700	560
800	16,000	1,600	1,067	800	640
900	18,000	1,800	1,200	900	720
1,000	20,000	2,000	1,333	1,000	800

EG: If the room has a total input of 200,000 Btuh, it requires a room volume of 4,000 cu ft. If the ceiling height is 15 feet, the area needed is 267 feet or a space roughly 16 x 16 feet.

Mbtuh = Input divided by 1,000

When using direct piped combustion air from the outside, there are two variables to consider. The O2 percentage of the air changes when the temperature changes. The warmer the air, the lower the O2 levels. The barometric conditions affect the air to fuel ratio also. The lower the barometric pressure, the lower the O2 percent in the air. If you look at the table below, you will see the boiler may be close to sooting as the combustion air temperature rises if adjusted at 0°F.

Effect of combustion air temperature on excess air

Combustion °F	Excess air %	Status
0°F *	15%	Safe
20°F	10%	Borderline safe
40°F	5%	Unsafe
60°F	-0.6%	Dangerous
80°F	-6%	Dangerous

* Combustion testing at this temperature

Ducted Combustion Air to Burner
Power Flame recommends sizing the combustion air duct for a pressure drop of 0.1" w.c. including all screens, filters, and fittings.

Combustion air piped to outside

Estimated Combustion Air Required @ Various Boiler Inputs

Boiler Btuh Input	Cu Ft Gas / Hr**	Cu Ft Gas/ Minute	CFM Air*
200,000	200	3.33	50
300,000	300	5.0	75
400,000	400	6.67	100
500,000	500	8.33	125
600,000	600	10.0	150
700,000	700	11.67	175
800,000	800	13.33	200
900,000	900	15.0	225
1,000,000	1,000	16.67	250
1,500,000	1,500	25.0	375
2,000,000	2,000	33.33	500
3,000,000	3,000	50.0	750

* Based upon 15 cubic feet of air for every cubic foot of gas burned.
** Based on 1,000 Btu per cubic foot of gas

Natural Gas

Average Btu Content of Common Fuels

Fuel Type	Number of Btu/ Unit
Fuel Oil #2	140,000 / Gallon
Fuel Oil #6	150,000 / Gallon
Butane	3,200 Btu's / CF
	21,500 Btus/ pound
	102,400 / Gallon
Natural Gas	1,025,000/ 1,000 cubic feet
Propane	91,330/ gallon
Coal	28,000,000 per ton
Electricity	3,412/ KWH
Wood Mixed	14,000,000/ cord or 3,500 / pound
Wood, Air Dried	20,000,000/ cord or 8,000 / pound
Kerosene	135,000 / gallon
Pellets	16,500,000/ton

Natural Gas Calculations

1 Cu ft Natural Gas =	1,000 Btus
1 MCF =	1,000,000 Btus
	1 MMBTU
	1 MCF
	1,000 Cu Ft.
	10 CCF
	10 Therms
1 Dekatherm	1 MCF
	10 Therms
	1,000,000 Btus
1 Therm =	100,000 Btus or 100 MBTU
	0.1 MCF
	1 CCF
	100 Cu. Feet
1 CCF =	1,000 Cu Ft
	100 Therm
Propane	
1 gallon =	92,000 Btus
1 Cu Foot =	2,250 Btus
#2 Fuel Oil	
1 Gallon =	140,000 Btus
Petroleum	
1 Barrel =	42 gallons

To get 1,000,000 Btu's you need the following	
Fuel Source	1,000,000 Btu's
Natural Gas @ 1000 Btu/ cu ft	1,000 Cu ft
Therms of natural gas	10
Dekatherm of natural gas	1
Coal @ 12,000 Btu/ lb.	83.333 Lb
Propane @ 91,600 Btu/ gal	10.917 Gal
Gasoline @ 125,000 Btus/gal	8.000 Gal
Fuel Oil #2 @ 140,000 Btus/gal	7.194 Gal
Fuel Oil #6 @ 150,000 Btus/gal	6.666 Gal
Electricity @ 3,412 Btu/kWh	293.083 Kwh
Boiler horsepower	29.31
Pounds of steam	1,000

Gas pipe sizing				
	Pipe Length			
Pipe Size	10 Feet	20 Feet	40 feet	80 Feet
	Capacity in Cubic Feet per hour			
½"	120	85	60	42
¾"	272	192	136	96
1"	547	387	273	193
1 1/4"	1,200	849	600	424
1 ½"	1,860	1,316	930	658
2"	3,759	2,658	1,880	1,330
2 ½"	6,169	4,362	3,084	2,189
4"	23,479	16,602	11,740	8,301

Maxitrol RV Series Gas Pressure Regulator		
Spring Color Ratings		
Spring Color	Inches W. C.	kPa
Plated	3-6"	0.75-1.5
Orange	4-8"	1-2
Blue	5-12"	1.25-3
Brown	1-3.5"	0.25-0.9
Plated	2-5"	0.5-1.25
Pink	3-8"	0.75-2
Violet	4-12"	1-3
Green	5-15"	1.25-3.7
Red	10-22"	2.5-5.5
Yellow	15-30"	3.7-7.5
Black	20-42"	5-10.5

Boiler Gas Consumption			
Btuh	Cu Feet/Hr	Cubic Feet/ Min	Cubic Feet/ Sec
1,000,000	1,000	16.67	0.28
2,000,000	2,000	33.33	0.56
3,000,000	3,000	50.00	0.83
4,000,000	4,000	66.67	1.11
5,000,000	5,000	83.33	1.39
6,000,000	6,000	100.00	1.67
7,000,000	7,000	116.67	1.94
8,000,000	8,000	133.33	2.22
9,000,000	9,000	150.00	2.50
10,000,000	10,000	166.67	2.78
11,000,000	11,000	183.33	3.06
12,000,000	12,000	200.00	3.33
13,000,000	13,000	216.67	3.61
14,000,000	14,000	233.33	3.89
15,000,000	15,000	250.00	4.17
16,000,000	16,000	266.67	4.44
17,000,000	17,000	283.33	4.72
18,000,000	18,000	300.00	5.00
19,000,000	19,000	316.67	5.28
20,000,000	20,000	333.33	5.56
Based upon 1,000 Btu's per cubic foot			

Gas Pressure Comparison					
Inches Hg	Ounces	PSI	Inches Hg	Ounces	PSI
0.1	0.05	0.003	13	7.51	0.468
0.2	0.12	0.007	14	8.09	0.504
0.4	0.23	0.01	15	8.67	0.54
0.6	0.35	0.02	16	9.24	0.576
0.8	0.46	0.028	17	9.82	0.612
1	0.58	0.036	18	10.4	0.648
2	1.15	0.072	19	10.98	0.684
3	1.73	0.108	20	11.56	0.72
4	2.31	0.144	21	12.13	0.756
5	2.89	0.18	22	12.71	0.792
6	3.46	0.216	23	13.29	0.828
7	4.04	0.252	24	13.87	0.864
8	4.62	0.288	25	14.45	0.9
9	5.20	0.324	26	15.02	0.936
10	5.78	0.36	27	15.60	0.972
11	6.35	0.396	28	16.18	1.008
12	6.93	0.432			

Convert mbar or millibar to inches W.C.

Millibar or mbar	Inches W.C.	Millibar or mbar	Inches W.C.
1	2.49	8	19.93
2	4.98	9	22.43
3	7.47	10	24.91
4	9.96	11	27.40
5	12.45	12	29.89
6	14.95	13	32.38
7	17.44	14	34.87

Convert pascals to inches of water column

$0.004014631332 \times \text{Pascals} = \text{Inches W.C.}$

Convert Inches W.C. to Pascals

$249.088875 \times \text{Inches W.C.} = \text{Pascals}$

Pascal	Inches H₂O	Pascal	Inches H₂O
1	0.00	130	0.52
5	0.02	135	0.54
10	0.04	140	0.56
15	0.06	145	0.58
20	0.08	150	0.60
25	0.10	155	0.62
30	0.12	160	0.64
35	0.14	165	0.66
40	0.16	170	0.68
45	0.18	175	0.70
50	0.20	180	0.72
55	0.22	185	0.74
60	0.24	190	0.76
65	0.26	195	0.78
70	0.28	200	0.80
75	0.30	205	0.82
80	0.32	210	0.84
85	0.34	215	0.86
90	0.36	220	0.88
95	0.38	225	0.90
100	0.40	230	0.92
105	0.42	235	0.94
110	0.44	240	0.96
115	0.46	245	0.98
120	0.48	250	1.00
125	0.50	255	1.02
		260	1.02

Convert Psi - Bar

Psi	Bar	Psi	Bar	Psi	Bar
1	0.07	21	1.45	41	2.83
2	0.14	22	1.52	42	2.9
3	0.21	23	1.59	43	2.97
4	0.28	24	1.65	44	3.03
5	0.34	25	1.72	45	3.1
6	0.41	26	1.79	46	3.17
7	0.48	27	1.86	47	3.24
8	0.55	28	1.93	48	3.31
9	0.62	29	2	49	3.38
10	0.69	30	2.07	50	3.45
11	0.76	31	2.14	51	3.52
12	0.83	32	2.21	52	3.59
13	0.9	33	2.28	53	3.65
14	0.97	34	2.34	54	3.72
15	1.03	35	2.41	55	3.79
16	1.1	36	2.48	56	3.86
17	1.17	37	2.55	57	3.93
18	1.24	38	2.62	58	4
19	1.31	39	2.69	59	4.07
20	1.38	40	2.76	60	4.14

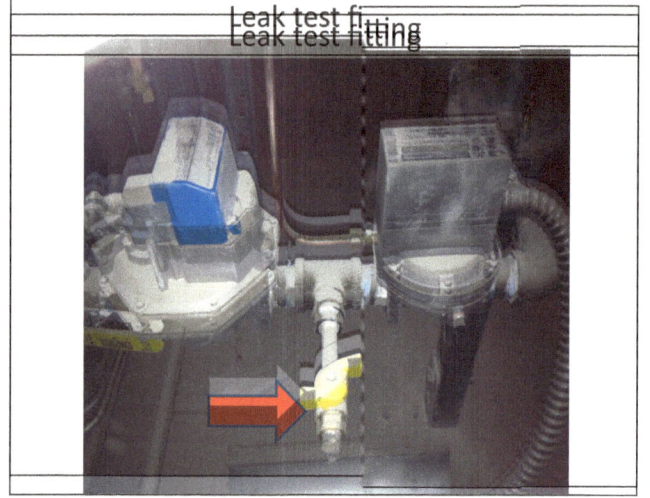

Leak test fitting

Leak testing gas valves.

The electric gas valves for the boiler should be tested yearly. Code allows a certain amount of acceptable leakage through a closed gas valve. The following are the recommendations provided by Honeywell for testing their gas valves.

How to leak test a gas valve

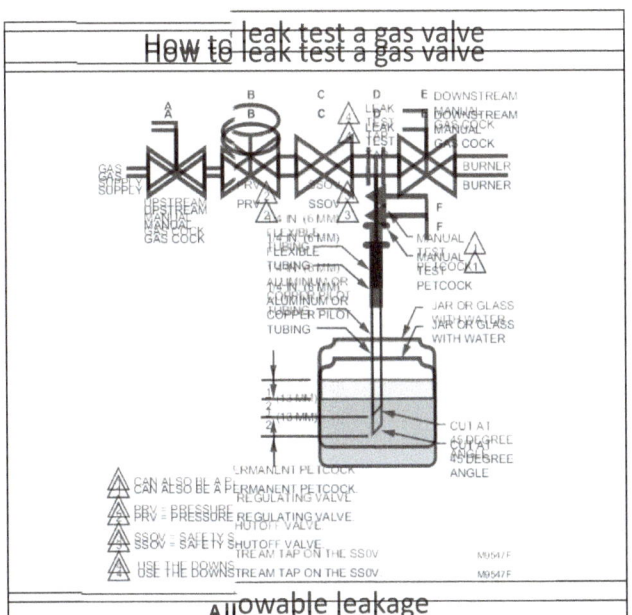

Allowable leakage

Table 5: Allowable Leakage for V48/V88 Valves.

V48/V88 Pipe Size (in.)	Air (cc/h)	Natural Gas (cc/h)[a]	Bubbles/10 sec., Max @ 45 degrees[b]
3/4	266	332.5	8
1	302	377.5	9
1-1/4	442	552.5	13
1-1/2	442	552.5	13
2	650	812.5	20
2-1/2	650	812.5	20
3	650	812.5	20

[a] Natural gas: multiply air by 1.25.
[b] Bubble leakage: Divide natural gas by 573, then multiply by 14.

> Draft controls are suggested when the stack is over 30 feet higher than a boiler in a Category 1 appliance.

Sizing Gas Train Manifold Vent
Source: Philadelphia Gas Works

Gas Train Venting

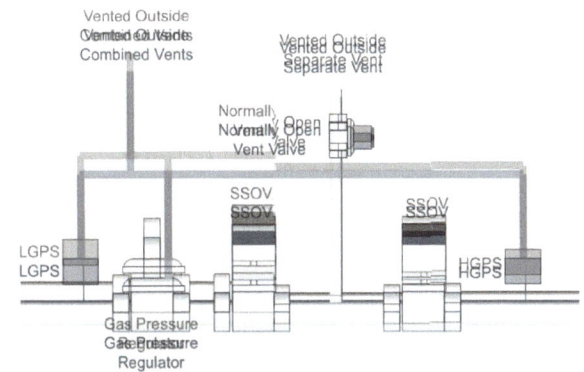

To calculate the common vent pipe size, the common pipe cross sectional area must be large enough for all the components. If we want to vent two regulators with 3/4" vents and two gas pressure switches with 1/4" vents. We will get our cross-sectional sizes from the chart below:

3/4" regulators   0.533 X 2 = 1.066
1/4" gas switches   0.104 X 2 = 0.208
                                    1.274

We would need a pipe size of 1 1/4" to combine all these vents.

- Pipe runs over 30 feet horizontal are not recommended and should be increased one pipe size for each 30 feet run.
- Normally open vent valves cannot be combined with other vents or other appliances.
- Vent terminations should be: 4 feet below, 1 foot above, and 3 feet horizontally from windows, doors, and gravity air intakes. 3 feet above any forced air inlet within 10 feet horizontally.
- Vents should have a screened 90 facing down.

Pipe Sizes		
Pipe Size	Inside Diameter	Inside Cross-Sectional Area
1/8"	0.269	0.057
1/4"	0.364	0.104
1/2"	0.622	0.304
3/4"	0.824	0.533
1"	1.049	0.864
1 1/4"	1.38	1.495
1 1/2"	1.61	2.036
2"	1.61	2.036
2 1/2"	2.469	4.788
3"	3.068	7.393
4"	4.026	12.73
5"	5.047	20.004
6"	6.065	28.89

**ASME CSD1 Sizing: If your locale follows the ASME CSD1 code, Section CF-190 has a different formula for sizing the manifold vent line. It requires "the manifolded line shall have a cross-sectional area not less than the area of the largest branch line directly piped to the manifolded line plus 50% of the additional cross-sectional areas of the manifolded branch lines.

Every cubic foot of gas burned produces the following:
- 8 cubic feet of nitrogen
- 2 cubic feet of water vapor
- 1 cubic foot of carbon dioxide

Fuel Oil

Oil Delivery Rate in Gallons per Hour					
Nozzle Rating	Pump Pressure				
	100	150	200	250	300
1.0	1.0	1.2	1.4	1.6	1.75
1.5	1.5	1.8	2.1	2.4	2.6
2.0	2.0	2.5	2.9	3.2	3.5
2.5	2.5	3.1	3.5	4.0	4.3
3.0	3.0	3.7	4.2	4.7	5.2
3.5	3.5	4.3	5.0	5.5	6.0
4.0	4.0	4.9	5.6	6.3	6.9
4.5	4.5	5.5	6.3	7.1	7.8
5.0	5.0	6.1	7.1	7.9	8.7
5.5	5.5	6.7	7.8	8.7	9.5
6.0	6.0	7.3	8.5	9.5	10.4
7.0	7.0	8.5	9.9	11.0	12.0
8.0	8.0	9.8	11.3	12.6	13.8
9.0	9.0	11.0	12.6	14.2	15.6
10.0	10.0	12.2	14.1	15.8	17.3
11.0	11.0	13.4	15.0	15.8	17.3
12.0	12.0	14.6	16.9	18.9	20.7
13.0	13.0	16.0	18.4	20.6	22.5
14.0	14.0	17.0	19.8	22.0	24.0
15.0	15.0	18.4	21.3	23.6	25.9

To calculate fuel oil flow rate:

$$\text{Flow} = \text{Nozzle GPH @ 100 Psi} \times \sqrt{\frac{Operating\ Pressure}{100}}$$

To calculate required pump pressure to get desired flow rate.

$$\text{Pump Pressure} = \left(\frac{Desired\ Flow\ Rate}{Nominal\ Flow\ @\ 100\ psi}\right)^2$$

Fuel Oil = 15% hydrogen and 85% Carbon

Allowable Tolerances for Oil Nozzles	
Flow Tolerances	Percentage of Tolerance
.30 to .45 GPH	+10% to 0%
.50 to .85 GPH	+5 to -3%
.85 to 1.50 GPH	+5% to -2%
1.65 GPH and above	+2.5% to -2.5%
Spray Angle Tolerances	
30° & 45°	+10% to 0%
60°, 70°, 80°, and 90°	+5% to 0%

Recommended Vacuum Readings for Oil Burners			
Single-Stage		Two-Stage	
5-10 in. vacuum, typ		10-15 in. vacuum, typ	
	Inches Mercury		Inches Mercury
1 pipe Capacity	< 6"	2 pipe	<17"
2 pipe	< 12"		

| Fuel Oil Tubing Size ||
Tube Size	Tubing Diameters
3/8"	.0086
½"	.00218
5/8"	.000785

Rule of Thumb for Vacuum Fuel Oil Systems	
1-inch vacuum =	One foot of vertical lift
	One 90⁰ Elbow
	Each 10 feet of horizontal suction pipe

Typical Oil Burner Properties		
Burner Type	Gal/Hr	Btuh
Vaporizing	0.15 – 2.5	20,000 to 350,000
Air Atomizing, Low Pressure	0.5-530	70,000 to 8,000,000
Air Atomizing, High Pressure	10-500	1,400,000 to 75,000,000
Steam Atomizing	10-500	1,400,000 to 75,000,000
Mechanical Atomizing, Non-recirculating	0.5-80	70,000 to 12,000,000
Mechanical Atomizing, Recirculating	25-1200	3,500,000 to 180,000,000

Nozzle Spray Patterns by Manufacturer			
	Solid	Hollow	Semi-Solid
Danfoss	AS	AH	AB
Delavan	B	A	SS
Hago	ES, P	H	SS
Monarch	R	NS, PL	PLP
Steinen	S	H	SS

Electrical

It's a good idea to check the tightness of electrical wires on a boiler. They loosen over time and could cause intermittent failures.

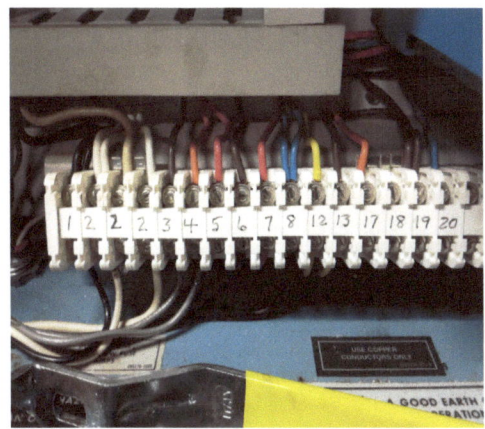

Electrical Equivalents Formulas	
Watt=	44.236 foot-pounds minute
	2,654.16 foot-pounds hour
	0.00134 hp
	3.414 Btu
	0.0035 lb. of water evaporated per hour
	44.236 foot-pounds minute
	2,654.16 foot-pounds hour
Kilowatt=	44,235 foot-pounds minute
	1.34 H.P.
	0.955 BTU per second
	57.3 Btu per minute
	3,413 Btuh per hour
	1,000 W
	1.34 hp
	3.53 lbs. water evaporated per hr from and at 212⁰F
1 H.P.=	33,000 foot-pounds minute
	746 watts
	42.746 Btu per minute
	2,564.76 Btu per hour
	0.746 kW
	33,475 Btuh
	772 ft lbs.
	550- ft-lb per second
1 Btu=	17.452 watts per minute
	0.2909 watts hour

Single Phase

To Find:	Single Phase
Amps when HP is known	$\dfrac{HP \times 746}{V \times \%EFF \times PF}$
Amps when KW is known	$\dfrac{KW \times 1{,}000}{V \times PF}$
Amps when kVA is known	$\dfrac{KvA \times 1{,}000}{V}$
Kilowatts	$\dfrac{V \times A \times PF}{1{,}000}$
HP	$\dfrac{V \times A \times PF \times \%Eff}{746}$

To Find:	Three Phase
Amps when HP is known	$\dfrac{HP \times 746}{V \times \%EFF \times PF \times 1.73}$
Amps when KW is known	$\dfrac{KW \times 1{,}000}{V \times PF \times 1.73}$
Amps when kVA is known	$\dfrac{KvA \times 1{,}000}{V \times 1.73}$
Kilowatts	$\dfrac{V \times A \times PF \times 1.73}{1{,}000}$
HP	$\dfrac{V \times A \times \%Eff \times PF \times 1.73}{746}$

Ohm's Law

Volts =	$\sqrt{Watts \times Ohms}$ Watts/Amperes Amps × Ohms
Amperes =	Volts/Ohms Watts/Volts $\sqrt{\dfrac{Watts}{Ohms}}$
Ohms =	Volts / Amperes Watts / $Amperes^2$ $Volts^2$ / Watts
Watts =	Volts × Amperes $Amperes^2$ × Ohms $Volts^2$ / Ohms

To find Amps, multiply watts × reciprocal of known electrical voltages

Voltage	Reciprocal
120	.008333
208	.004807
230	.004347
277	.003610
480	.002083

Example: If the unit is rated for 5,000 watts and 208 volts, the amps will be 24.03

Wire Size & Amp Ratings Copper

	140°F	167°F	194°F
Wire Types	NM-B, UF-B	THW, THWN, SE, USE, XHHW	THWN-2, THHN, XHHW-2
Wire Gauge		Amp ratings	
14	15	20	25
12	20	25	30
10	30	35	40
8	40	50	55
6	55	65	75
4	70	85	95
3	85	100	115
2	95	115	130
1		130	145
1/0		150	170
2/0		175	195
3/0		200	225
4/0		230	260

Full Load Amperes of Single-Phase Motors

HP	RPM	115V	230V
1/8	1725	2.8	1.4
	1140	3.4	1.7
	860	4.0	2.0
1/4	1725	4.6	2.3
	1140	6.15	3.07
	860	7.5	3.75
1/3	1725	5.2	2.6
	1140	6.25	3.13
	860	7.35	3.67
1/2	1725	7.4	3.7
	1140	9.15	4.57
	860	12.8	6.4
3/4	1725	10.2	5.1
	1140	12.5	6.25
	860	15.1	7.55
1	1725	13.0	6.5
	1140	15.1	7.55
	860	15.9	7.95

Full Load Amperes of Three Phase Motors

HP	RPM	115V	230V
1/4	1725	0.95	0.48
	1140	1.4	0.7
	860	1.6	0.8
1/3	1725	1.19	0.6
	1140	1.59	0.8

	860	1.8	0.9
½	1725	1.72	0.86
	1140	2.15	1.08
	860	2.38	1.19
¾	1725	2.40	1.23
	1140	2.92	1.46
	860	3.28	1.63
1	1725	3.16	1.58
	1140	3.7	1.85
	860	4.12	2.06
1 ½	1725	4.61	2.31
	1140	5.18	2.59
	860	5.75	2.88
2	1725	5.98	2.99
	1140	6.28	3.14
	860	7.28	3.64
3	1725	8.70	4.35
	1140	9.25	4.62
	860	10.3	5.15
5	1725	14.0	7.0
	1140	14.6	7.3
	860	16.2	8.1
7 ½	1725	20.3	10.2
	1140	20.9	10.5
	860	23.0	11.5

Color Code of NB-B Romex Cable

The following is the color code of the cable so you can quickly tell the gauge of the wire.

Jacket Color	Wire Size
White	14 AWG
Yellow	12 AWG
Orange	10 AWG
Black	8 AWG
Black	6 AWG

Fuse Sizing Guide

Class CC Time Delay

Multiply Motor FLA from nameplate x 2. Round up or down to nearest fuse size. Example:
1 HP @ 115 Volt = 15.1 Full Load Amps
15.1 x 2 = 30.2 Use 30 Amp fuse

Class J Time Delay

Multiply Motor FLA from nameplate x 1.5. Round up or down to nearest fuse size. Example:
½ HP @ 115 Volt = 7.4 Full Load Amps
7.4 x 1.5 = 11.10 Use 12 Amp fuse

Average Conduit Wire Capacity
Wire Type: THHN/THWN/THWN-2

Conduit Type EMT

Wire Size	½"	¾"	1"	1 ½"	2"
#14	12	22	35	84	138
#12	9	16	25	61	101
#10	5	10	16	38	63
#8	3	6	9	22	36
#6	1	4	6	16	26
#4	1	2	4	9	16
#3	1	1	3	8	13
#2	1	1	2	7	11
#1	0	1	1	5	8

Conduit Type ENT

Wire Size	½"	¾"	1"	1 ½"	2"
#14	10	18	32	80	132
#12	7	13	23	58	96
#10	4	8	14	36	60
#8	2	5	8	21	34
#6	1	3	6	15	25
#4	1	1	3	9	15
#3	1	1	3	7	13
#2	1	1	2	6	11
#1	0	1	1	4	8

Conduit Type Flex

Wire Size	3/8"	½"	¾"	1"	1 ½"	2"
#14	4	13	22	33	76	135
#12	3	9	16	24	55	98
#10	1	5	10	15	35	61
#8	1	3	5	8	20	35
#6	1	1	4	6	14	25
#4	0	1	2	3	9	15
#3	0	1	1	3	7	13
#2	0	1	1	2	6	11
#1	0	0	1	1	4	8

Conduit Type IMC

Wire Size	½"	¾"	1"	1 ½"	2"
#14	14	24	39	91	150
#12	10	17	28	67	109
#10	6	11	18	42	68
#8	3	6	10	24	39
#6	2	4	7	17	28
#4	1	2	4	10	17
#3	1	1	3	9	14

	#2	1	1	3	7	12
	#1	1	1	1	5	9

Conduit Type LFMC					
Wire Size	Conduit Sizes					
	3/8"	½"	¾"	1"	1 ½"	2"
#14	7	12	22	36	81	134
#12	5	9	16	26	59	97
#10	3	5	10	16	37	61
#8	1	3	5	9	21	35
#6	1	1	4	6	15	25
#4	1	1	2	4	9	15
#3	1	1	1	3	8	13
#2	0	1	1	3	6	11
#1	0	1	1	1	5	8

Conduit Type PVC 40						
Wire Size	Conduit Sizes				
	½"	¾"	1"	1 ½"	2"
#14	11	20	34	82	136
#12	8	15	25	59	99
#10	5	9	15	37	62
#8	3	5	9	21	35
#6	1	4	6	15	25
#4	1	1	4	9	15
#3	1	1	3	8	13
#2	1	1	2	6	11
#1	0	1	1	5	8

Standard 24 Volt Thermostat Connections		
Terminal	Usage	Wire colors
R or V	24 V power	Red
Rh or 4	24 V Heating Power	Red
Rc	24 V Cooling Power	Red
C	24 V Common	Black
Y	1st Stage Cooling	Yellow
Y2	2nd Stage Cooling	Blue or Orange
W	1st Stage Heat	White
W2	2nd Stage Heat	No Standard Color
G	Fan	Green

Venting the Boilers

7 Times Rule - The flow area of the largest common vent or stack shall not exceed seven times the area of the smallest draft hood outlet.

Horizontal vs. Vertical - The horizontal vent (L) must be no more than 75% of the vertical height (H) of the flue. If using B Vent, the horizontal length can be the same as the height of the chimney, as per International Fuel Gas Code, 2006 503.10.9.

7 Times Rule Round Flue			
Smallest Draft Hood Outlet	Largest Common Flue	Smallest Draft Hood Outlet	Largest Common Flue
3"	7"	7"	18"
4"	10"	8"	22"
5"	12"	9"	22"
6"	14"	10"	26"

7 Times Rule Rectangular Flue		
Smallest Draft Hood Outlet	Area of outlet	Largest Common Vent Area
3"	7.06"	49"
4"	12.56"	88"
5"	19.64"	137"
6"	28.27"	198"
7"	38.48"	269"
8"	50.27"	352"
10"	78.54"	550"

Appliance Venting Categories			
Cat	Condensing	Efficiency	Pos/Neg
I	Non-condensing	83% or less	Negative
II	Condensing	Over 83%	Negative
III	Non-Condensing	83% or less	Positive
IV	Condensing	Over 83%	Positive

> Hydronic distribution is 15 times more efficient than refrigerant distribution.

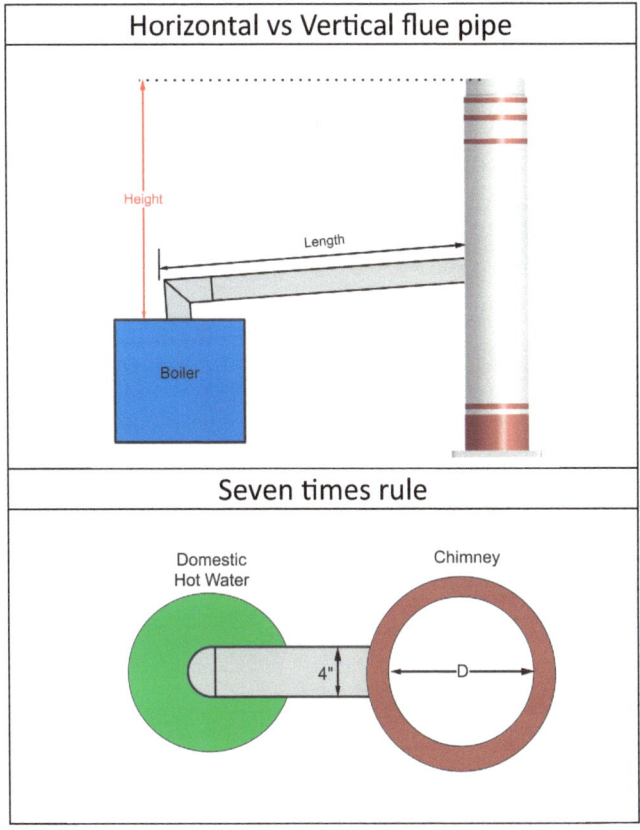

Sidewall Venting

If venting the boiler through the sidewall, the following are the International Mechanical Codes covering the installation:

Where adjacent to walkways, the termination of mechanical draft systems shall not be less than 7 feet above walkway.

3 feet above any forced air inlet within 10 feet

4 feet below, 4 feet horizontally from or 1 foot above any door, window, or gravity vent into building

No closer than 3 feet from an interior corner formed by 2 walls perpendicular to each other.

Not within 3 feet horizontally or directly above an oil tank or gas meter

At least 12 inches above finished grade.

Will You Need a Draft Control?
The project may require a draft control if:

- The new boiler is a Category 1 (Non-Condensing) appliance and will be vented into the existing chimney.
- The existing chimney is over 30 feet high.

What type of draft control?
Barometric dampers are more commonly used.

What type of barometric damper?
Oil - Single acting.
Gas or Gas/Oil - Double acting

When using a double acting damper on a boiler which fires only gas, the red stops should be removed. Many municipalities and codes require a spill switch to be installed if using a double acting barometric damper. The spill switch will sense flue gas spillage and shut off the boiler.

Where should the barometric damper be installed? According to Fields Controls, a leading manufacturer of barometric controls, the suggest the following locations in order of preference, A being the best:

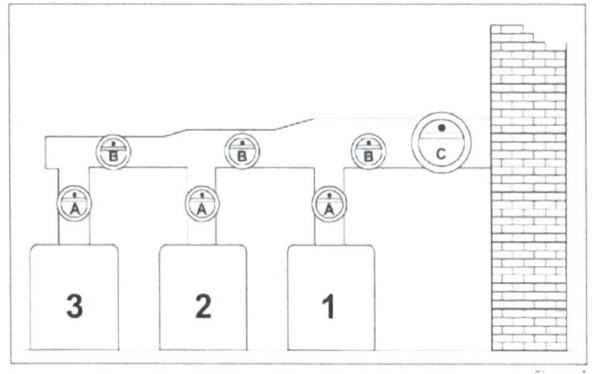

Preferred Barometric Damper Locations

Will you need a chimney liner?

If the new boilers do not use the existing chimney and a water heater is still vented into the chimney, the water heater may need a chimney liner. The "7 Times Rule," says, the flow area of the largest common vent or stack shall not exceed seven times the area of the smallest draft hood outlet. If the chimney size is more than the seven times rule, a liner must be used for the water heater flue.

Flue Information

Typical Vent Temperature Ranges

Venting Material	Temperature Ratings	Fuel
AL 29-4C Stainless	0 - 480° F	Gas
B and BW Vent	0 - 550° F	Gas
L Vent	0 - 1,000° F	Oil
Factory Built Chimney	500° - 2,200° F	Oil/Gas
Masonry Chimney	360° - 1,800° F	Oil/Gas
Verify with manufacturer		

Minimum Flue Gas Temperatures for Category 1 Boilers

Fuel	Minimum Flue Temperature for Non-Condensing Appliances
Natural Gas	265° F plus 1/2° F for each foot of stack or breeching, including horizontal and vertical runs
#2 Fuel Oil	240° F plus 1/2° F for each foot of stack or breeching, including horizontal and vertical runs

Acid Rain and Stack Temperature

Fuel	Dew Point Temperature	Minimum Stack Temperature
Natural Gas	150	250
#2 Fuel Oil	180	275
Non-Condensing Appliances		

Typical Boiler Exhaust Velocity in Feet per Second

Equipment Type	Exhaust Velocity ft/s
On Off Burner	16-26
Two step burner	31-49
Modulating burner	49-82
Minimum to keep surface free from soot	9.8-13

Typical Draft Readings for Boilers

Type of Heating System	Overfire Draft In. W.C.	Stack Draft In. W.C.
Gas, Atmospheric	N/A	-.02 to -.04
Gas, Power Burner	-.02"	-.02 to -.04
Oil, Conventional	-.02"	-.04 to -.06
Oil, Flame Retention	-.02"	-.04 to -.06
Positive Overfire Oil & Gas	+.4 to +.6	-.02 to -.04
Category 4 Positive	Positive	±1.0

Theoretical Chimney Draft @ 60° F

Average Chimney Temperature	Chimney Height in Ft		
	10	20	30
100	0.01	0.02	0.03
200	0.03	0.06	0.09
300	0.05	0.09	0.14
400	0.06	0.12	0.17
500	0.07	0.13	0.20
600	0.07	0.15	0.22
700	0.08	0.16	0.24
800	0.09	0.17	0.26
900	0.09	0.18	0.27

Theoretical Chimney Draft @ 0° F

Average Chimney Temperature	Chimney Height		
	10	20	30
100	0.03	0.05	0.08
200	0.04	0.09	0.13
300	0.06	0.12	0.17
400	0.07	0.14	0.20
500	0.08	0.15	0.23
600	0.08	0.17	0.25
700	0.09	0.18	0.27
800	0.09	0.19	0.28
900	0.10	0.19	0.29

Conversion Factors

Pressure Unit Conversions

Divide the known number by the desired pressure number to get desired pressure.

Known	Desired Pressure	
	Pounds Per sq. In.	Ounces Per Sq. In.
Centimeters of Water	0.0981	0.227
Feet of water	0.433	6.94
Inches Mercury	0.491	7.86
Inches Water	0.0361	0.578
Ounces per Square Inch	0.0625	---------
Pounds per Sq. Inch	---------	16.0

Known	Inches of Water	Feet of Water
Centimeters of Water	0.384	0.0328
Feet of water	12.0	0.883
Inches Mercury	13.6	1.13
Inches Water	---------	0.0833
Ounces per Square Inch	1.73	0.144
Pounds per Sq. Inch	27.7	2.31

Diameter to Circumference

Diam	Circum	Diam	Circum
12	37.70	26	87.96
14	43.98	30	94.25
16	50.27	32	100.53
18	56.55	34	106.81
20	62.83	36	113.10
22	69.15	38	119.38
24	75.40	40	125.66
26	81.68		

Sheet Metal Gauge

Thickness in decimals of an inch

Gauge Number	US Standard Gage for Uncoated Hot & Cold Rolled Sheets	Galvanized Sheet Gage for Hot-Dipped Zinc Coated Sheets
3	0.2391	
4	0.2242	
5	0.2092	
6	0.1943	
7	0.1793	
8	0.1644	0.1681
9	0.1495	0.1532
10	0.1345	0.1382
11	0.1196	0.1233
12	0.1046	0.1084
13	0.0897	0.0934
14	0.0747	0.0785
15	0.0673	0.071
16	0.0598	0.0635
17	0.0538	0.0575
18	0.0478	0.0516
19	0.0418	0.0456
20	0.0359	0.0396
21	0.0329	0.0366
22	0.0299	0.0336
23	0.0269	0.0306
24	0.0239	0.0276
25	0.0209	0.0247
26	0.0179	0.0217
27	0.0164	0.0202
28	0.0149	0.0187

Circular Equivalent of Duct

Circular Equivalent of Rectangular Duct Inches

Width	Height in Inches							
	6	8	10	12	14	16	18	20
6	7	8	8	9	10	10	11	11
8	8	9	10	11	11	12	13	13
10	8	10	11	12	13	14	15	15
12	9	11	12	13	14	15	16	17
14	10	11	13	14	15	16	17	18
16	10	12	14	15	16	17	19	20
18	11	13	15	16	17	19	20	21
20	11	13	15	17	18	20	21	22
22	12	14	16	18	19	20	22	23
24	12	15	17	18	20	21	23	24
26	13	15	17	19	21	22	24	25
28	13	16	18	20	21	23	24	26
30	14	16	18	20	22	24	25	27
36	15	17	20	22	24	26	27	29
42	16	19	21	23	26	28	29	31
48	17	20	22	25	27	29	31	33

Circular Equivalent of Rectangular Duct Inches

Width	Height in Inches							
	22	24	26	28	30	36	42	48
6	12	12	13	13	14	15	16	17
8	14	15	15	16	16	15	19	20
10	16	17	17	18	18	17	21	22
12	18	18	19	20	20	20	23	25
14	19	20	21	21	22	22	26	27
16	20	21	22	23	24	24	28	29
18	23	23	24	24	26	26	29	31
20	24	24	25	26	27	27	31	33
22	25	25	26	27	28	29	33	35
24	26	26	27	28	29	31	34	37
26	27	27	28	29	31	32	36	38
28	28	28	29	31	32	35	37	40
30	31	29	31	32	33	36	39	41
36	33	32	33	35	36	39	42	45
42	35	34	36	37	39	42	46	49
48	37	37	38	40	41	45	49	52

Rectangular Tank Capacity in Gallons
Length x Width x Height divided by 231 = Gallons

Rectangular Tank Height 12"

	US Gallons			
	Width			
Length	12	18	24	36
12	7	11	15	22
18	11	17	22	34
24	15	22	30	45
30	19	28	37	56
36	22	34	45	67
42	26	39	52	79
48	30	45	60	90
60	37	56	75	112

Rectangular Tank Height 18"

	Width			
Length	12	18	24	36
12	11	17	22	34
18	17	25	34	50
24	22	34	45	67
30	28	42	56	84
36	34	50	67	101
42	39	59	79	118
48	45	67	90	135
60	56	84	112	168

Rectangular Tank Height 24"

	Width			
Length	12	18	24	36
12	15	22	30	45
18	22	34	45	67
24	30	45	60	90
30	37	56	75	112
36	45	67	90	135
42	52	79	105	157
48	60	90	120	180
60	75	112	150	224

Rectangular Tank Height 36"

	Width			
Length	12	18	24	36
12	22	34	45	67
18	34	50	67	101
24	45	67	90	135
30	56	84	112	168
36	67	101	135	202
42	79	118	157	236
48	90	135	180	269
60	112	168	224	337

Circular Storage Tank Capacity in Gallons
Multiply ½ Tank Diameter by itself.
Multiply that by 3.146 x Length of Tank in inches.
Divide by 231 = Gallons of water

Circular tanks					
Estimate Storage Tank Capacity in Gallons					
Length (feet)	US Gallons				
	18	24	30	36	42
1	1.1	1.96	3.06	4.41	5.99
2	26	47	73	105	144
2.5	33	59	91	131	180
3	40	71	100	158	216
3.5	46	83	129	184	252
4	53	95	147	210	288
4.5	59	107	165	238	324
5	66	119	181	264	360
5.5	73	130	201	290	396
6	79	141	219	315	432
6.5	88	155	236	340	468
7	92	165	255	368	504
7.5	99	179	278	396	540
8	106	190	291	423	576
9	119	212	330	476	648
10	132	236	366	529	720
12	157	282	440	634	864
14	185	329	514	740	1008
	US Gallons				
Length (feet)	Inside Diameter (Inches)				
	48	54	60	66	72
1	7.83	9.91	12.24	14.41	17.62
2	188	238	294	356	423
2.5	235	298	367	445	530
3	282	357	440	534	635
3.5	329	416	513	623	740
4	376	475	586	712	846
4.5	423	534	660	800	852
5	470	596	734	899	1057
5.5	517	655	808	978	1163
6	564	714	880	1066	1268
6.5	611	770	954	1156	1374
7	658	832	1028	1244	1480
7.5	705	889	1101	1355	1586
8	752	949	1175	1424	1691
9	846	1071	1322	1599	1903
10	940	1189	1463	1780	2114
12	1128	1428	1762	2133	2537
14	1316	1666	2056	2490	2960

Length & Area	
1 Mile =	1,760 yards
	5,280 feet
	63,360 inches
	1.609 Km
1 Foot =	0.3048 M
	30.48 Cm
	304.8 mm
1 Inch =	25,400 microns
	2.54 CM
	25.4 Mm
1 acre =	43,560 Sq. Ft
	4,840 Sq. Yds
	0.4047 Hectares
1 League =	3.0 Miles
1 Fathom =	6 feet
	1.828804 meters
1 Furlong	660 feet
1 Sq. Mile =	640 Acres
1 Sq. Yd =	9 Sq. Ft
	1,296 Sq. Inches
1 Sq. Foot =	144 Square Inches
1 Cu Yard =	27 Cu Ft
	46,656 cu inches
	1,616 pints
	807.9 quarts
	764.6 Liters
1 Cu foot =	1,728 cubic inches
Diameter of Circle =	Circumference x 0.3188
Circle Circumference =	Diameter x 3.1416
Liquid	
1 Gallon =	8.33 pounds
	4 quarts
	8 pints
	3.785 liters
	0.13368 Cu Feet
	231 Cu Inches
1 Liter =	0.2642 Gallons
	1.057 quarts
	2,113 pints
Miscellaneous	
1 Barrel Oil =	42 gallons
Hours in a Year	8,760

Speed

1 MPH =	5,280 ft / hr
	88 ft/min
	1.467 ft/sec
	0.868 Knots per Hr
1 Knot =	1.1515 MPH

Pressure

1 lb. steam =	1 lb. H2O
14.7 psi =	33.95 ft H2O
	29.92 in Hg
	407.2 in w.g
	2,116.8 lbs./ Sq. Ft
1 Psia =	Psig = 14.7
1 psi =	2.307 Ft H2O
	2.036 in Hg
	16 ounces
	27.7 in w.c.
1 ounce =	1.73 inches w.c.
	0.4335 psi
1 Ft H2O =	0.4335 psi
	62.43 lbs./ sq. feet

Weight

# =	Pounds of Pressure
1 Lb =	16 oz
	7,000 grains
	0.4536 Kg
1 ton =	2,000 lbs.
	907 Kg

Liquid Measurements

1 Teaspoon =	60 drops
1 Tablespoon =	3 teaspoons
1 Ounce =	2 tablespoons
1 Measuring cup =	16 tablespoons or 8 ounces
1 Pint =	2 cups
1 Quart =	2 pints
1 Gallon =	4 quarts

Common Fraction to Decimal to Millimeters

Fraction	Decimal	Millimeters
1/16	0.0625	1.587
1/8	0.125	3.175
3/16	0.1875	4.762
1/4	0.250	6.350
5/16	0.3125	7.937
3/8	0.375	9.525
7/16	0.4375	11.113
1/2	0.50	12.700
9/16	0.5625	0.5625
5/8	0.625	15.875
11/16	0.6875	17.462
3/4	0.750	19.050
13/16	0.8125	20.637
7/8	0.875	22.225
15/16	0.9375	23.812
1	1.00	25.400

> A hydronic relief valve will lift slowly & a steam pop safety opens quickly.

Metric Conversions

1 kcal =	3.968 BTU
1 kg =	0.00930 BTU Pounds x 0.4536
1 KJ/Hr =	Btuh x 1.055
1 KW =	0.948 BTU/sec.
1 kcal/kg =	1.80 BTU/lb.
1 kcal/m^3 =	0.112 BTU/cu. ft
1 kcal/m^{2}h =	0.369 BTU/sq.ft.h
1 kcal/m^{2}h$^\circ$C =	0.205 BTU/sq.ft.h$^\circ$F
1 kcal/mh$^\circ$C =	0.67 BTU/ft.h$^\circ$F
1 kcal m/h$^\circ$C m^2 =	8.07 BTU in/sq.ft.g$^\circ$F
1 kcal/kg$^\circ$C =	0.999 BTU/lb.$^\circ$F
1 kcal/m^3 $^\circ$C =	0.0624 BTU/cu ft $^\circ$F
KJ/Hr =	Btu/h x 1.055
CMM =	CFM x 0.02832
LPM =	GPM x 3.785
KJ/Lb. =	Btu/Lb x 2.326
Meters =	Feet x 0.3048
Sq. Meters =	Sq. Feet x 0.0929
Cu. Meters =	Cu. Feet x 0.02832
Kg =	Pounds x 0.4536
Kg/Cu. Meter =	Pounds. Cu Feet x 16.017
Cu. Meters/ Kg =	Cu. Ft/ Pound x 0.0624
1 BTU =	0.252 kcal 107.7 kgm
1 BTU/sec. =	1.055 KW
1 BTU/lb. =	0.5556 kcal/kg
1 BTU/cu ft. =	8.900 kcal/m3
1 BTU/sq.ft.h =	2.71 kcal/m^{2}h
1 BTU/sq.ft.h$^\circ$F =	4.886 kcal/m^{2}h$^\circ$C
	1.49 kcal/m h$^\circ$C
1 BTU in/sq.ft.hr$^\circ$F =	0.124 kcal/mh$^\circ$C
1 BTU/lb.$^\circ$F =	1.001 kcal/kg$^\circ$C
1 BTU cu. ft$^\circ$F =	16.2 kcal/m3 $^\circ$C

Metric Conversion Factors

Multiply	By	To Obtain
Inches	2.54	Centimeters
Feet of Water	2.24165	Centimeters of Mercury
Ft^3	28.32	Liters
In^3	16.39	Cm^3
Gallons	3.7853	Liters
GPM	0.06308	Liters/Sec
Btu/Hr	0.252	Kilocalories/Hr
PSI	0.0703	Kilogram/ Cm^3
Liters	0.26418	Gallons
Liters	61.025	Cubic In
Liters	0.0353	Cu. Feet
Liters	0.001	Cu. Meters
Liters	2.202	Lbs water
Cu. Meters	264.17	Gallons
Cu. Meters	999.972	Liters
Cu. Meters	61023.74	Cu. Inches
Cu. Meters	35.3145	Cu. Feet
Cu. Meters	2202.61	Lbs Water

Convert Temperature Readings

Fahrenheit to Celsius	Celsius to Fahrenheit
$\dfrac{\text{Degrees F} - 32}{1.8}$	$(1.8 * \text{Degrees C}) + 32$
Celsius to Kelvin	Kelvin to Celsius
K = C + 273.15	C = K − 273.15

Fahrenheit to Celsius Temperatures

$^\circ$F	$^\circ$C	$^\circ$F	$^\circ$C	$^\circ$F	$^\circ$C
0	-17.8	11	-11.7	40	4.4
1	-17.2	12	-11.1	50	10.0
2	-16.7	13	-10.6	60	15.6
3	-16.1	14	-10.0	70	21.1
4	-15.6	15	-9.4	80	26.7
5	-15.0	16	-8.9	90	32.2
6	-14.4	17	-8.3	100	37.8
7	-13.9	18	-7.8	150	65.6
8	-13.3	19	-7.2	200	93.3
9	-12.8	20	-6.7	212	100.0
10	-12.2	32	0.0	215	101.6

Energy Unit Conversions			
	kg.m.	kcal	Joule
kg.m.	1	0.00235	9.81
kcal	427	1	4186.8
Joule	0.102	0.000239	1
KW / hr	3,67,098	860	3.600×10^6
HP hour	2,73,745	641.18	26,84,519
	kW hour		HP hour
kg.m.	0.27×10^{-5}		0.365×10^{-5}
kcal	0.001161		0.001556
Joule	27.77×10^{-8}		37.25×10^{-8}
KW hour	1		1.3411
HP hour	0.74565		1

Metric Pressure	
6.8947 kPa	1 pound per sq. in (psi)
9.794 kPa	1m column of water
1 kPa	10.2 cm of water
1.3332 kPa	1 cm column of water
3.3864 kPa	1 inch of mercury (in Hg)
8 kPa	6 cm of mercury
1kg/cm^2	14.223 lbs./sq. in Psi
1 mm WG	0.002927 in mercury
0.0703kg/cm^2	1 lb./sq. in. Psi
340.39 mm WG	1 in. Mercury
43.9 mm WG	1 ounce/sq. inch
6.9 bar	1 psig
0.45 kg	1 pound

Metric Liquid	
Metric	U.S.
3.7854 L	1 Gallon
0.946 L	1 Quart
0.473 L	1 Pint
1 L	0.264 Gallons
1 L	33.814 Ounces
29.576 ml	1 Fluid Ounce
236.584 ml	1 Cup

Metric Length & Area	
Metric	U.S.
1m	39.37 inches
	3.28 feet
	1.094 yards
	0.0016 miles
1m^2	10.764 Ft2
1.609 km	1 mile
25.4 mm	1 inch

2.54 cm	1 inch
304.8 mm	1 foot
1 mm	0.03937 inches
1 cm	0.3937 inches
1cm2	0.1550 in2
1 dm	3.937 inches

Common Metric Abbreviations	
Prefixes/Symbols	
Giga – G	1,000,000,000
Mega – M	1,000,000
Kilo -k	1,000
Hecto – h	100
Deka/deca -da	10
Deci -d	1/10
Centi – c	1/100
Milli – m	1/1,000
Micro - µ	1/1,000,000
Nano – n	1/1,000,000,000

Metric Abbreviations	
Abbreviation	Name
A-C	
A	Amp
C	Celsius
C	Coulomb
cg	Centigram
cl	Centiliter
cm	Centimeter
cm^2	Cubic Centimeter
dag	Dekagram
dal	Decaliter
dam	Decameter
dg	Decigram
dm	Decimeter
dl	Deciliter
dm^2	Cubic Decimeter
g	Gram
GW	Gigawatt
hg	Hectogram
hl	Hectoliter
hm	Hectometer
K	Kelvin
kcal	Kilocalorie
kcal/m^3	Kilocalorie/Cubic Meter
kg	Kilogram
kj	Kilojoule
kjh	Kilojoule/Hour

kl	Kiloliter
km	Kilometer
Km²	Square Kilometer
km/h	Kilometer per hour
kPa	Kilopascal
kW	Kilowatt
kWh	Kilowatt – hour
l	Liter
m	Meter
mA	Milliamp
mAh	Milliamp-hour
mg	Milligram
Mj	Megajoule
ml	Milliliter
mm	Millimeter
m/s	Meter per second
V	Volt
W	Watt

Definitions:

Air Change: The amount of air required to completely replace the air in the boiler and associated flue passages.

Air, Primary: Air mixed with the fuel to provide combustion.

Air, Secondary: Air mixed with the flue gases to provide proper turbulence to allow complete combustion.

Air shutter: A device to control the airflow to the burner.

Air, Tertiary: Air from the boiler room introduced to the flue to overcome excessive chimney draft. It is sometimes called Dilution Air.

Air to Fuel Ratio: This is the amount of air used in the combustion process. It is typically 15 parts of air for each part of natural gas.

Barometric Damper: A damper installed in the flue piping to control the excessive draft in a Category 1 boiler by introducing boiler room air.

Boiler: A closed vessel to heat water or create steam.

Boiler, Cast Iron: A boiler, which uses cast iron as its heat exchanger.

Boiler, Condensing: A boiler that operates below the flue gas dew point temperature.

Boiler, Steel: A boiler, which uses steel as its heat exchanger.

Boiler, Fire Tube: A boiler where the flue gases travel through the tubes.

Boiler, Water Tube: A boiler where water flows through the tubes.

Boiler, Modular: A heating system consisting of several smaller boilers.

Breeching: A conduit to transport the combustion by-products from the boiler to the outside or to the chimney. It is also called a flue.

Btu (British Thermal Unit): The amount of heat required to raise one pound of water, one-degree F.

Btuh: Btus in one hour

Burner: A mechanical device which mixes air and fuel to provide ignition and combustion of the fuel.

Burner, Atmospheric: A burner that uses natural draft and gas pressure to provide combustion.

Burner, Power: A burner that uses an internal blower to mix the fuel and the air for combustion.

Carbon Dioxide / CO2: A gas produced as a by-product of combustion.

Carbon Monoxide /CO: This deadly gas is odorless and tasteless. It's produced when the combustion is out of adjustment.

Combustion Air: The air introduced from the outside required for the proper combustion of the fuel.

Combustion Analyzer: A device which measures the flue gas and efficiency of the boiler.

Control, Operating: A device that senses water temperature & starts or stops the burner. This is usually set for 180°F.

Control, Limit: A device that starts or stops the burner. This is usually set for a higher pressure than the operating control. In most applications, it is a manual reset.

Delta T: Temperature difference.

Dew Point Temperature: The flue gas temperature which is cooled enough to allow the water vapor to condense into water.

Differential Pressure: The pressure difference across the inlet and outlet.

Dirt Leg: Nipples and a pipe cap installed just before the gas train to capture any dirt in the gas line before it enters the gas train.

Draft: The pressure differential between atmospheric pressure and the pressure in the flue and boiler.

Draft Diverter: A device above the boiler where tertiary air is introduced to the flue after the main combustion.

Draft, Mechanical: The pressure differential between atmospheric pressure and the pressure in the flue and boiler that is induced because of a fan or blower.

Draft, Natural: The pressure differential between atmospheric pressure and the pressure in the flue and boiler without a fan or blower.

Emergency Door Switch: This switch is installed just inside or outside the boiler room exits. When pushed, it will shut off all the boilers.

Firing Rate: The burning rate of fuel and air in the burner.

Firing Rate Control: A control that senses the boiler temperature. It will regulate the burner between low and high fire to meet the desired set point. It's also called the modulating control.

Flue: A conduit that transports the combustion by-products from the boiler to the outside or to the chimney, also called breeching.

Flue Gases: The by-products of combustion produced by the burner & measured in boiler flue.

Fuel Train: The gas pressure regulator, nipples, and gas valves, located in the gas piping directly attached to the burner, also called a gas train.

Gas Pressure Regulator: A device that controls the gas pressure supplied to the burner.

Gas Pressure Switch: A safety device that senses the available gas pressure and will shut the boiler off in the event the pressure is outside of the setting.

Heat, Latent: The amount of heat required to cause a change of state.

Heat, Sensible: The amount of heat required to cause a change in temperature.

Heating Medium: The material that the boiler heats. It could be steam, water, or some other type of fluid.

High Fire: The highest design firing rate of the burner.

Lag Boiler: The boiler is not the first boiler to start when there is a call for heat.

Lead Boiler: The boiler is the first boiler to start on a call for heat.

Lockout: A safety shutdown that requires a manual reset of the control or safety device.

Low Fire: This is the lowest design firing rate of the burner.

Low Fire Start: This switch verifies that the burner is in the "Low Fire" position before opening fuel valves.

Low High Low Fire: A burner that starts at low fire and will travel between low & high while there is a call for heat.

Low High Off Fire: A burner that starts at low fire and goes to high fire. The burner will stay at high fire until the call for heat ends.

Low Water Cutoff, Primary: A device that senses the water level inside the boiler and will shut down the burner if the water level drops to an unsafe level.

Low Water Cutoff, Auxiliary: This shuts off the burner if the water levels drop below its setpoint. This one has a manual reset switch, and the boiler will not restart until the water level is restored and the reset button is pushed.

Modulating Burner: A burner that will operate at any position from low to high fire to meet the demands of the modulating control.

Modulating Control: A control that senses the water temperature. It will set the burner at any position from low to high fire.

Non-Condensing Boiler: A boiler with flue gas temperature above the dew point temperature.

Pilot, Continuous: It is a pilot flame that burns all the time, regardless of whether the burner is firing.

Pilot, Intermittent. It is a pilot that lights when there is a call for heat. The pilot will stay light during the entire time the main burner is firing.

Pilot, Interrupted: It is a pilot that lights when there is a call for heat. The pilot will shut off once the main flame is established.

Pipe Pitch: The amount of slope used to avoid water pooling. It is typically 1 inch slope away from the boiler every ten feet.

PPM: Part per Million.

Prepurge: On a call for heat, the burner blower starts to purge the boiler combustion chamber and flue passages of any unburnt fuels. It will operate long enough to provide several air changes inside the boiler.

Pressure Control, Operating: This pressure control will be used to cycle the boiler on and off.

Pressure Control, Limit: This pressure control is a manual reset control, set at a higher pressure than the operating control. The boiler will not restart until the boiler pressure drops below the setpoint of the control and the manual reset button is pushed.

Relief Valve: A valve located on a boiler that will open gradually to relieve the excess pressure inside the boiler if the pressure rises to the setpoint of the valve.

Sidewall Venting: Boiler flue that is piped to the side wall of the building rather than a chimney or stack.

Spill Switch: A device located by a draft diverter or a barometric damper that senses rollout of the flue gases and shuts off the burner. It is a manual reset switch.

Static Head: This is the weight at the bottom of a column of pipe.

Strainer: This has an internal screen used to prevent dirt, rust, and mud from entering.

Total Dissolved Solids: These are primarily inorganic substances found in in water such as salts, magnesium, calcium, potassium, sodium, nitrates, chlorides, & sulfites. These are often referred to as TDS.

WC: This stands for Water Column

A

Air Venting Tips I Learned, 37

B

Baseboard radiation is ticking, 24
Bearing assembly is leaking, 14
Boiler draft is too high, 9
Boiler draft is too low, 9
Boiler draft is wrong, 9
Boiler makes a moaning sound when firing, 22
Boiler noises, 22
Boiler pressure keeps rising, 9
Boiler Room Safety, 4
Boiler Temperature Controls, 18
Burner blower starts but flame doesn't, 8
Burner short cycles, 8

C

Chattering in pipes, 22
Circular Equivalent of Duct, 80
Combustion, 58
Combustion Air, 66
Condensing boiler is not condensing, 11
Conversion Factors, 81
Corrosion in system, 15

D

Degree Days and Design Temperatures, 42
Discoloration of boiler jacket, 9

E

Entire building is cool, 13
Estimate Hydronic System Voloume, 52
Excessive air in system, 14
Excessive chemical treatment use, 9
Excessive combustion noise, 23
Excessive vibration in system, 24
Expansion Compression Tank Sizing, 47

F

Flame rolls from under boiler base, 10
Flue Information, 78
Fuel Oil, 72

G

Glycol, 53

H

Heating Formulas, 45
Heating Water with Electric, 40
High fuel costs, 10
How to tell if expansion tank is flooded, 17
Hydronic cast iron radiator ratings, 41
Hydronic Formulas, 46
Hydronic Piping, 25

I

Intermittent Flame Failures, 7

L

Leak testing gas valves., 71

M

Main air vent is squealing / noisy, 23
Manual reset limit control is tripping, 12
Metric, 85

N

No heat from baseboard radiation, 15
No heat to AHU coil, 14

O

One area overheating, 13
Outdoor Design Temperatures, 45

P

Pilot lights but main flame doesn't, 7
Pipe Insulation, 57
Pipes are banging, 23
Piping, 54
Piping makes a gurgling noise, 22
Pump is grinding / banging, 23
Pumps, 49

R

Radiator is dripping, 15
Radiator not heating, 14
Relief valve is opening, 17
Rust under draft diverter, 10

S

Sizing a Circulator, 49
Sizing Gas Train Manifold Vent, 71
Some parts of the building are cold, 13
Squealing noise when boiler runs, 22
System is losing water, 15

T

The boiler is not firing, 6
The boiler is off on low water, 8
The boiler is sooted, 10
The boiler makes a hissing sound, 22
The boiler makes a humming or buzzing sound, 22
The boiler makes a ticking/cracking noise when firing, 22
The expansion tank keeps flooding, 16
The flue gas temperature is higher, 11
The pump is squealing, 22
The temperature rise through the boiler is too high, 11
Top floor cool while bottom floors are warm, 14
Troubleshooting hydronic piping, 29
Trouble-shooting Hydronic Systems, 31
Typical Combustion Readings for Category I boilers, 59

U

Unbalanced heat in building, 14
Understanding the Boiler Sequence of Operation, 39

V

Velocity Calculation, 56

W

Water Density, 51
Water Formulas, 50
Water under boiler, 9, 21

Z

Zoning with Pumps, 34
Zoning with Valves, 35

www.ingramcontent.com/pod-product-compliance
Lightning Source LLC
Chambersburg PA
CBHW051915210526
45473CB00006B/2017